Create Meaning with

Create Meaning with Sound: Sound Design Theory for Live Theatre offers an artistic framework that connects philosophy, perception, and storytelling to the practice of theatrical sound design, mapping out a theory of sound design for the performing arts.

Drawing from history – from the amphitheaters of ancient Greece to the modern digitally-networked stage – the book traces how humans have used sound to shape emotion, convey truth, and deepen narrative experience. This work provides clear definitions of fundamental concepts such as intention, choice, and meaning, establishing a shared language for designers and collaborators alike. Moving beyond technical manuals, each chapter combines theory with practical reflection and creative exercises that encourage readers to develop their own artistic voice.

Designed for students, educators, and practitioners of Sound Design and Theatre Design, *Create Meaning with Sound* binds art and science to redefine sound design as a creative and interpretive art form in its own right. This book serves as both a classroom text and a creative reference, inviting readers to see themselves not merely as technicians but as storytellers who use sound to inspire emotion, provoke thought, and create meaning on stage and beyond.

Tony Angelini is a sound designer, author, podcaster, and creativity coach. With more than 25 years of experience in live theatre and over 300 productions to his credit both in the USA and internationally, he brings an artistic and human-centered approach to sound and storytelling. His work explores the intersection of creativity, technology, and the performing arts. He has composed two albums (*Breathing Deeply* and *Bold Girls*) and a book of poetry (*Walking the Creeper*), and he has contributed articles to *Pro Audio Review Magazine*.

Create Meaning with Sound

Sound Design Theory for Live Theatre

Tony Angelini

Routledge
Taylor & Francis Group
NEW YORK AND LONDON

Designed cover image: Tony Angelini

First published 2027
by Routledge
605 Third Avenue, New York, NY 10158

and by Routledge
4 Park Square, Milton Park, Abingdon, Oxon, OX14 4RN

Routledge is an imprint of the Taylor & Francis Group, an informa business

Library of Congress Cataloging-in-Publication Data
Names: Angelini, Tony, author
Title: Create meaning with sound : sound design theory for live theatre / Tony Angelini.
Description: New York, NY : Routledge, 2027. | Includes bibliographical references and index.
Identifiers: LCCN 2026010291 (print) | LCCN 2026010292 (ebook) | ISBN 9781032856797 hardback | ISBN 9781032856780 paperback | ISBN 9781003519393 ebook
Subjects: LCSH: Theaters—Sound effects | Theaters—Production and direction
Classification: LCC PN2091.S6 A64 2027 (print) | LCC PN2091.S6 (ebook) | DDC 792.02/4—dc23/eng/20260522
LC record available at https://lccn.loc.gov/2026010291
LC ebook record available at https://lccn.loc.gov/2026010292

ISBN: 978-1-032-85679-7 (hbk)
ISBN: 978-1-032-85678-0 (pbk)
ISBN: 978-1-003-51939-3 (ebk)

DOI: 10.4324/9781003519393

Typeset in Palatino
by Apex CoVantage, LLC

Dedication

Dedicated to Sharon, who grew up with a theatre dad yet still managed to turn out pretty great.

Contents

Foreword

Allow me, please, to indulge just a bit, so that I may accurately give you a true and well-deserved picture of Tony Angelini. I will start by saying that in my 47 years as a director, in New York as well as two very prestigious LORT regional theatres, and finally 27 years as the Producing Artistic Director of Barter Theatre, I can say without hesitation that Tony Angelini is the most creative and most forward thinking of any of the sound designers I have ever had the pleasure to work with. In short, Tony is the most intelligent, passionate sound designer I have ever known.

To give some brief credentials to my reference above, I have worked professionally as a sound designer for theatre since the age of 18, put myself through graduate school at the University of California, Davis, as the sound designer for the theatre department there and, in New York, worked with and was good friends with a sound designer, who has since become a five-time Tony Award winner on Broadway. And yet, I still hold Tony as a superior designer especially for the brilliant definitions, deeper understanding, and impact that make Tony's sound designs unique, the process and creation of which he fully explains beautifully within this book.

The most important aspect of Tony's theory is that he begins to use and define sound design as a richer, more important, more impactful, and more integrated part of the theatre than has been true in the past and is only now becoming an essential element of today's theatre experience. The goal here is that the artistry of the sound designer is every bit as significant to a show as is each actor upon the stage, as is the costume designer, as is the set designer, as is the lighting designer, as is every artist involved in the stage production all working to tell the story first and foremost, not to stand out alone, but to work together as artists to create an experience for the audience, which fulfills the meaning and the enjoyment and the purpose of the production.

Turning to the beginnings of live theatre presentations, sound effects were created mechanically by offstage personnel for each and every performance. This was true with live theatre for more than 2,000 years (according to Britannica). Thus, it is no wonder that sound is largely thought of as a technical creation. Following the age of radio, which developed the ability to record sound effects and the creation of sound effect libraries, the prerecorded materials were largely used by theatre sound technicians. Thus, every baby crying, every dog barking, every clap of thunder, etc., sounded largely the same regardless of the theatre production. Add the use of microphones, followed by highly advanced sound boards and sound equipment, the world of sound was and is largely technologically based.

Even in the current day, the world of theatre sound is more based on volume of music, hearing the performers speak, accurately chosen sound effects, and occasionally wild, surprising, original, creative effects. Thus, sound today is still largely about the technology of the equipment and the clever usage of the equipment by the sound engineer and not the artistic purpose of the production.

Time and time again, in discussions about theatre, inevitably, the question arises, "What is the purpose of a Sound Designer?" When you turn that question back to the person asking, they generally say that the person running the sound board controls the volume of the sound and basically the mix of the sound and plays the recorded sound effects. But they have really no understanding of what "sound design" means and thus why a sound designer even exists. Because audiences largely believe that sound exists simply to help them hear the show better.

What Tony Angelini does brilliantly in this book is not only to address the purpose and the necessity of what the sound designer is and should be, but makes clear as to why it is essential that the sound designer understands their role as an artist who works in tandem and as a collaborator with all of the other artists creating the show whose goal is to make the production as rich, layered, meaningful, and entertaining as possible.

Tony has given us a new road to help the sound designer use technology combined with artistry to establish a greater use of sound for the improvement of today and future theatre productions.

– Richard Rose

Richard Rose served for 27 years as Producing Artistic Director for historic Barter Theatre, the State Theatre of Virginia, until his retirement in 2019. During his time at Barter, Rick more than quadrupled Barter's annual attendance, with over 163,000 people attending Barter annually. He grew Barter's income from $747,000 in annual revenues prior to his arrival to $9 million annually. Under Rick's leadership, Barter won numerous business awards being recognized throughout Virginia, nationally, and internationally. While at Barter, Rick directed, choreographed, designed, and/or wrote over 125 productions. Original adaptations, to name a few, written by Rick include *A Modern Christmas Carol*. *Dracula!*, *Wuthering Heights*, *The Hound of the Baskervilles*, and, acknowledged in American Theatre Magazine as one of the best new regional theatre plays of the year, *Frankenstein*. Rick has his Master of Fine Arts in directing and design from the University of California at Davis, where he also served as the sound designer for the university's theatre department and held the position of producing artistic director for the campus' musical theatre program. He has a bachelor of arts from St. Norbert College, DePere, Wisconsin, where he was honored with a "distinguished achievement award in Humanities." Rick did postgraduate work, directed at New York University, and worked for the prestigious Juilliard School in New York City.

He has directed and produced numerous shows in New York and at many regional theatres throughout the USA.

Most recently, Rick directed and produced the world premiere and the Southern Tour of the new American musical *We'll Meet Again*, which he helped develop through Barter's Appalachian Festival of Plays and Playwrights, a program he created at Barter to help Appalachian and Southern plays and playwrights achieve national recognition. Rick also helped found the Theatre Department at Emory & Henry College.

Preface

The thought of writing a book about this topic came from a colleague of mine during a telephone conversation late one night. She posed the simple and straightforward question, "What exactly is Sound Design?" As I began to explain, it quickly became apparent to me that no two sound designers would have the same answer to that question. I decided we require a common understanding, a concise definition, and I was as good as anyone to begin that conversation. I hope the conversation persists because when we put our heads together, we become far more than we are alone.

Acknowledgments

Moonwalking with Einstein: The Art and Science of Remembering Everything by Joshua Foer (Penguin Press, 2011) – A quotation from this work is used by permission of the author.

Man's Search for Meaning by Viktor E. Frankl (Beacon Press, 1959) is cited for its enduring insight into purpose and human experience. A quotation from this work is used by permission of Beacon Press, Boston.

The 7 Habits of Highly Effective People by Stephen R. Covey (Free Press, 1989) – A quotation from this work is used by permission of FranklinCovey Co.

The Predictive Mind by Jakob Hohwy (Oxford University Press, 2013) informs the discussion of perception, prediction, and cognition in relation to sound and meaning. A quotation from this work is used by fair use.

How Emotions Are Made: The Secret Life of the Brain by Lisa Feldman Barrett (HarperCollins, 2017) – A quotation from this work is used by permission of HarperCollins Publishers.

Finally, I would like to acknowledge Tom Bilyeu and Impact Theory University (Impact Theory LLC, 2017–present), along with Delia Peterson, Arame Ndoye, Katie Moran, M.J. Coppola, and Steven Schrembeck for their life-changing impact for which I am truly grateful.

Figures by Chapter

Chapter 4: From Sound to Meaning

Chapter 7: The Sound Designer's Technique

Chapter 8: The Sound Designer's Process

Chapter 9: Early and Late Classical Period

Chapter 11: The Renaissance – Transitioning to the Modern Theatre

Chapter 12: The Italian Renaissance – The Birth of Opera

Introduction

This is an artistic approach to sound design for theatre and live events. Many of the ideas in this book will cross over to other artistic disciplines besides sound design, such as film, podcasts, and games.

This book, at its core, is a book about philosophy. The ideas explored in this book are not mentioned in books about audio engineering and system design. Most of the books available regarding sound design do not explore these topics in any usable depth. They tend to put the cart before the horse. Building your sound system (what I call the delivery system) and then deciding what it will be used for after it is built is like jumping out of an airplane and then building your parachute once you land. The steps to the plan have to be re-ordered if you are going to make a successful landing. In other words, it is better to first know what you want to present artistically, and then you can build the delivery system and use it in an artful way.

When I began to write this book, I had very different intentions about what it would be. I was going to write about things like the physics of sound, how the brain interprets sound, and sound system design and engineering. I was going to write about things related to a sound designer's career, such as how to get work, production meetings, what happens in technical rehearsals, and things like that. Yet, as I began writing about all of that, I realized that those topics were not exactly what I wanted to communicate: Topics like those already have sufficient clarity, and a person can learn about them through a plethora of books, through examples, with mentoring, and with practice. (Practice is crucial to learning and to mastering.)

DOI: 10.4324/9781003519393-1

The more I wrote, the more I realized that what I really wanted to give to the reader was a sense of clarity of the big picture. "High order" ideas that have been heretofore ambiguous. Things that we all "kind of" know intuitively but have not exactly put into words. I know in my core that if these ideas are made concrete, then people can take them, personalize them, and use them to create great art of their own.

It was not always clear to decide how much practical detail should be included. Entering the rabbit hole of too much information might distract the readers from the main topic of focus. I always kept the purpose of the book in mind, which was to write about the concepts of the theory, and leave the execution of said theory to the reader.

So, you would not find much about sound system engineering, how to record and edit a sound sample, or how to interact with a stage manager so that they call a sound cue at just the right time. What you will find is an approach to sound design. A way of thinking about it that will help you understand, imagine, and create great art using the medium of sound. It is an old profession, older than any of the other technical arts associated with theatre, yet it is rarely considered an art form of its own.

Sound designers are artists, and as artists, I hope many will weigh in on this sound design theory. I hope it sparks discussion, which will shine a light on where I might have got it wrong, and where I got it right. After all, art is growth. You grow when you learn, practice, and create.

Basic Artistic Practices

Some of the principles I have learned from this exercise of writing a book were life-changing to me, and they boil down to:

1. Commit to the daily practice of your art.
   If you want to get fit and decide to go to the gym for the first time and work out for 6 hours all at once, you will be disappointed. If you do that, you will probably do more harm to your body than good. However, if you go to the gym every day and have a sincere workout for only 20 minutes each day, then over time, you are going to be in the best shape you have ever been.
2. Enjoy the process.
   Enjoy the minutes, hours, and days you work on your art. This is both a choice and not a choice. Be aware of how you are feeling as you work. If you enjoy it, recognize that you do, and that will keep you going on the inevitable occasions when it becomes overwhelm-

ing. Everyone has days of flow when good energy is high, and you can work for hours, while other days drag on with little or no energy. There are ways to choose to enjoy even the hardest moments. At those moments, I remind myself how good I will feel when I am done. The only way to get past it is to go through it. I have always felt that working as a sound designer in the theatre, my worst days were better than most people's highest holidays. It is a belief that gives me energy to persevere during those hard days.

3. Detach from the outcome.
   As I said, the book that I intended to write is very different from how it turned out. Being open to letting it become what it wanted to become resulted in a work I could not have expected or hoped for. Do not get me wrong, intentions are everything. You must begin the creation process with a specific intention of your own choosing, or you will not be able to start at all. Then, as you create and remain open to what you learn from the process, you may find that your intentions change. You will find that it becomes more than the sum of you. It will take on a life of its own, as they say. The only way to achieve that is to let it go and let it grow.

Reorient Your Perspective

When you are building your show, details matter. Whether you are setting your DAW plug-ins, creating patch sheets for your delivery system, and tuning your public address (PA) system or setting the perfect EQ curve on an actor's lavalier microphone, your perspective must necessarily become very narrow. Your focus must be on the granularity of that task. However, before you choose your tools or even select your preliminary intentions for the show, I encourage you to pull back your perspective. Ask and know the bigger questions. That is what this book seeks to ask. Why am I creating with sound? What is my purpose in this profession? What is my duty? What is my purpose for this specific project? What is my intention for specific moments in this project and how are those intentions informed by my higher personal goals for this profession? What tools, technology, and resources are available to me in order to manifest those intentions?

Parts of this book will be precise regarding the technique of designing, and other parts will be about a less precise methodology that will guide you to think for yourself and express those thoughts that are derived from the best parts of you as an individual and unique voice. Whether you know it or not, you already have an individual "artistic voice" inside you. If you have not become acquainted with that voice yet, this book will help you to get to know it.

I also encourage you to engage in the conversation. Talk with other sound designers. Talk with me. Talk with other artists in the theatre. Talk with other artists outside of the theatre. Learn as much as you can from anybody you can, including from yourself.

Create. I can't wait to see what you come up with.

– Tony Angelini

Part 1

Sound Design Theory

Part 1 is dedicated to sound design theory itself. It encourages artists to use this foundation to express themselves through sound in their own way. This theory provides a structure that can be a foundation for infinite paths of aural expression.

For the purposes of this discussion, there is a balance to be achieved between presenting philosophy and presenting the details of the practice of that philosophy – tools, technology, current standards of professional practices, etc. Judicial consideration of how much practical detail to be included in this discussion had to be exercised. The criterion for making those judgments was simple – did those details – such as how to set up a sound system for a production, how to set up gain structure for lavalier microphones, how to create a signal flow diagram, for example – clarify and codify the specific philosophy being discussed, or did they obscure it. Hopefully, that balance has been struck.

DOI: 10.4324/9781003519393-2

1

Common Language

It was a hot afternoon in London on this fateful day in the summer of 1613 and probably humid in a theatre so close to the River Thames. The actor portraying King Henry VIII was standing backstage of The Globe, waiting to enter. It was the last scene of Act 1 of William Shakespeare's new(-ish) play, *All is True*. He probably did not know that in ten years, the *First Folio* would include this play for posterity with the new title *King Henry VIII*. Interestingly enough, this actor was quite possibly involved in the construction of that portfolio, maybe even renaming the play after the main character, the very part that he portrayed. Ironically, this actor's actual name has been lost to history.

As this was only the second or third performance of the run of the show, he was probably running line changes in his head while he waited for his entrance. Sweat dripped from his temple and down the inside of his costume. Typically, any changes from the playwright would have been given to him earlier that morning. As for his entrance cue, it would be hard to miss. This entrance cue was the firing of a cannon that could be heard two furlongs away. All of the ordnance for battles and other armament sound effects were exploded from behind the theatre, on the ground. Gunpowder was expensive, but this was the favorite theatre of Queen Elizabeth's court, so it had to have the finest production values money could buy.

The moment came, the cannon was fired, and the actor entered. There would have been no actual cannonball fired, only powder. There is no telling if this actor saw the bit of fire that flew awry and landed on the theatre's

DOI: 10.4324/9781003519393-3

thatched roof, but it rapidly set the whole theatre ablaze. A few people were injured while as many as 3,000 people fled the theatre. Fortunately, no one perished in the fire, according to the many written accounts of the incident.

The Globe Theatre burned completely to the ground. The Globe was no more. History gone, for a time, until construction of the new building quickly began. In the meantime, the theatre was dark and so losing money every single day.

Shakespeare's famous theatre, The Globe, was destroyed by a sound technician. Sound design is an ancient and proud profession, indeed.

366 years later, Walter Murch was the first to be credited as a sound designer and is generally credited as coining the title for the film *Apocalypse Now* in 1979. He wanted to legitimize the craft, comparing it to the roles of production designer and costume designer. These days, sound designers work not only on film projects but also in the game industry, the podcast industry, the recording industry, and even radio drama.

In this book, we are going to focus on sound design for theatre and live events. For the rest of this book, whenever the term sound designer is used, it will be in reference to theatrical sound design.

What Exactly Does a Sound Designer Do?

There seem to be many roles a sound designer assumes when working on a live production. So much so that the function and purpose of sound design proper can become obscured and invariably misunderstood. Everyone seems to "sort of know" what we do, but we are far from a common definition. Our area of responsibility varies from production project to production project. In one production, the sound designer may be expected to plan and create sound effects, while in another production, that same sound designer may be expected to compose music. In yet another, they may be hired to install the sound system and mix the show. To be fair, these roles depend on factors like theatre budgets and personnel availability, but mostly, it is the producers who are unclear. Producers have budgets and need to make a profit in order to keep producing, and we, as sound designers, need to remember that. It is up to us to educate producers and other theatre personnel on what our role is and what we are capable of when we can focus all our creative energy into that role.

Even sound designers do not seem to agree on what we do as a profession, not exactly and not universally. How can producers and other colleagues keep it straight if we as sound designers cannot seem to agree on what we do and how we go about doing it?

A common language is needed regarding what we do, how we do it, and why we do it. Without a clear definition of what sound design is and its purpose, the role of the sound designer becomes overwhelmingly confusing. There is a standards-and-practices vacuum where sound design is concerned.

In this book, we are also going to explore a theory of sound design for the live theatre. It will define the profession, the craft, establish fundamental standards, and develop a common language that will serve as a framework that all sound designers can stand on to guide choices and inform decisions.

A theory is simply a supposition, or a system of ideas intended to explain something. It is a set of principles on which the practice of an activity is based. A principle is a fundamental truth or proposition that serves as the foundation for a system of belief or behavior or for a chain of reasoning.

That is what sound design theory is. A set of principles that serve as a foundation that we as sound designers can use and from which we can grow in our craft as artists.

This theory of sound design will provide direction. It will increase artistic impact. It will allow the sound designer to start to create. It will help the sound designer start from a clear beginning and not the muddy middle.

You may be looking for a book that explains the physics of sound or sound system engineering, but that is not what this book is about. Technology changes over time. Our understanding of how the universe works changes. This book posits something more fundamental than that. It is a way of looking at our art and living your art. A way of thinking that can guide you in your journey of learning, experiencing, growing, and mastering that art. I cannot wait to see what you will do with it.

Chapter 1 Exercises

Journaling Challenge: Get comfortable making choices of your own.

If you could change the title of sound designer to anything you wanted, what would you call yourself? Same Duties. Same Function. Same Roles. Pick a name.

The first duty of an artist is to make choices based on what you think, what you like, and what you have learned from experience, from other people, or from your own reasoning and ideas. In other words, an artist makes choices based on their individual aesthetic.

I challenge you to make your first choice as a sound designer-artist. What are you going to call what you do?

When someone asks me what I do, or what is my role in a given production process, I just tell them that I am the sound guy. That satisfies them.

Everyone knows what the sound guy does, right? For most people, it is hard to explain completely, but they kind of know. I call myself the sound guy because I am responsible for the sound, music, and sound system for the show, and I am a guy. I am the sound guy.

Yet, that is not exactly comprehensive. Depending on the needs of the show, I am the composer, or the recording engineer, or the audio production editor, or if it is a situation where I am leading a team (usually the case), and I might be none of the above. As a team leader, I may be something like a sound producer or sound director. In the past, I have been the mixing engineer and sound designer at the same time. Sometimes, I am the sound design teacher or sound design mentor. Sometimes, I am the show control programmer, and sometimes, I am part producer or even audio editor.

Journal on It

What is a name that can efficiently explain all of that? All of the varied tasks and roles mentioned earlier still do not fully explain the definition of a sound designer.

If you could coin a title to encompass everything you can, might, and will do as a sound designer, what would it be? Imagine everyone will use it. All of the other designers on the team, the actors, the books, all of the show programs, all of the websites, now and for posterity, would call it what you have named it. There might even be a category added to the Tony Awards with the name that you come up with.

You may be wondering what is the point of this little exercise. It is because I want you to get used to making choices based on something originating from you. It should come from that place in yourself where creative ideas are born. Make a decision based on your own impetus, arrogance, love, thought, and heart. That is why I am asking you to do this. Seriously, I would like to hear what you come up with.

Now, do not ask anyone else. This needs to be all you. Do not Google it. Do not check a book. Do not ask your best friend. Do not talk about it, except to yourself. Out loud, if that is part of your creative process. Think it through, for a minute or ten or 60.

The title you give this occupation is going to be based on a combination of your current understanding of what a sound designer does, along with some imagination, and a pinch of your intention. When you get the right mix (no pun intended), you will know it. It will just feel good to you. Keep working on it based on what you learn and think.

This title is allowed to change as you practice your art and learn and grow. The title is allowed to evolve as you evolve and as your understanding of the occupation evolves.

If your understanding is limited and you do not think you are qualified for this, do not worry. You are 100% qualified to make a choice that is based on where you are at this moment, what you currently know, and how you currently feel.

When creating art, a wrong choice is better than no choice at all.

Here is a secret: There are no wrong choices, and there are only choices that do not serve your vision. Even then, it is a matter of comparison of which will get you there most efficiently. Which choice will have the most impact aligned with your intention?

So begin your journaling on this. Then keep journaling for the length of your career.

2

Foundational Concepts in Sound Design Theory

For the sake of clarity and for the sake of a common understanding, let us agree on the meaning of a few important terms. I am going to boldly define them here in this chapter so that we can then move on to more complex concepts. These definitions are for our purpose of sound design theory.

Some of these terms have been debated for millennia, and we may never pin down their full complexity. Nuances change as people change and as culture evolves. They are defined here in this context and this context only. This way, everyone knows exactly what we are talking about without getting bogged down on the basics. Otherwise, we will never make it to the higher ideas. There will be plenty of time for further conversation about these topics, and I will welcome it.

Sound Design Theory

Sound design theory is a system of ideas based on a collected set of fundamental principles. It is the framework of a method of using those principles to create your design.

It serves to standardize an understanding of the sound designer's art among the various colleagues and artists you may work with.

It is a foundation for the sound designer. If you master these principles and put them into practice, you will be able to achieve higher and more complex meaning in your work.

DOI: 10.4324/9781003519393-4

Sound Design Defined

Sound design, broken down into the most basic components, has two definitions:

1. The act of planning what the audience will hear and how they will hear it.
2. The plan of what the audience will hear and how they will hear it.

Sound design can be the act of planning or the plan itself, much like the word painting can mean the act of painting a work of art or the completed work of art itself.

The first definition is simple enough. The act of planning what the audience will hear, along with how it will be delivered to the audience. For example, let us say an actor is to deliver a monologue alone on the stage. Let us say it will be in a big venue. The highest goal of sound reinforcement is to ensure that every member of the audience receives that dialogue with intelligibility. Another way to put that is that the highest priority of the sound designer is to make sure every member of the audience understands every word (or lyric). If you cannot hear the monologue, you cannot follow the story. Therefore, the actor's words fall into the category of what the audience will hear. Congratulations, you have made your first design decision.

Creating a Plan Means Making Your Best Choices

This first definition also includes creating a plan on how the audience will hear it. This means making some choices. Is the actor's unprocessed voice going to reach all the way to the back row, or do you need to plan to reinforce the voice with a delivery system that includes a microphone, amplifiers, carefully placed speakers, and all of the ancillary gear that goes with it?

Do you think it would tell a better story if the amplified voice had effects such as reverb, echo, or dynamic processing? Do you think there should be sound effects played during the monologue? What about music? As a sound designer, you should keep all options on the table. Anything is fair game if it contributes to the storytelling.

Your choices are only limited by the constraints that the production puts upon you, such as space, budget, and personnel. Other than that, the sky is the limit. No, not the sky: the universe.

The First Duty

Consider it your first duty to be creative in a way that only you can be. Bring the best of yourself as you work within the confines of the constraints mentioned earlier. Make your choices authentic. Personalize it. Make personal choices based on what fascinates you or energizes you.

Make your choices holistic. Bring your whole self to your choices – everything you have learned and experienced in your life, your personal tastes and predilections, your ideas, your joy and sadness. They are all a part of you. Trust that it is all inside you, and it all informs your best decision. Trust your instincts. Please yourself.

The Plan Itself

The second definition cites that a sound design is the plan itself. You have made choices and collected them into a cogent plan. That plan is your design. It is yours. It exists.

Here, we come to an important distinction: the design exists in concept only. It exists in your mind, and in the minds of people you share it with. The written description is not the plan. The cue sheet is not the plan. The mixing engineer's script is not the plan, nor is any other paperwork, schematics, or signal flow diagrams. The paperwork only serves these purposes:

- Memory aides for you and your crew.
- Communication media so that others can understand the plan.
- Reference material for archival purposes.

Paperwork is not the plan itself, any more than a book is the story it tells. This is an important distinction because it will inform how you create those memory aides and communication media, and why they are useful.

Some designers, myself included, combine the decision-making process with the building process. For example, it is my process to build the delivery system (sound system) paperwork to help me piece it together, thinking and designing as I draw. I also do that with the cue sheet and creation of the recorded sound effects. Although I use them in tandem, I never confuse the decision-making process with the tools that I may use to help me make those decisions.

Your plan is never static. Consider your plan a draft. It can be changed as you learn, grow, and as the production evolves. As the actors and directors do their work to build the show, you can and should be available to adapt and grow with that process. I never consider my plan static. My designs are always a work in progress. I am at liberty to alter the plan or to throw it out whole cloth and start over at any time. After all, the design is not me – the design is my work. I may love something I have created, but I never forget this axiom: *Sound serves the show, the show doesn't serve the sound.*

Even though on opening night, I make the declaration that no changes happen after opening night, that declaration is more for the performers and crew than for me. The design becomes a stable, predictable aspect to the show that the actors and crew can depend on, but as for me, if I did not think it was best for the show to cease changes and updates, I might be making tweaks and improvements until the show closes. Letting go is a choice and, therefore, part of the design plan itself. The design process is only complete when you, the designer, decide that you are finished. Even then, the design process is never finished. It is you who will decide when you are finished working on it.

The Purpose of Sound Design

The answer to the question "What is the purpose of sound design?" is complex. There are many levels to the question, so here, we will deconstruct the answer starting from the granular purpose and expand to the larger purposes.

On the most fundamental level, the purpose of any sound design is to raise the audience's level of understanding of the story that you and the playwright, director, designers, actors, and other collaborators are telling. The sound informs the audience of information they need to know, such as period, place, mood, texture, and feeling. This is done on a micro level (a specific moment in a scene) and on a macro level (the overall arc of the story and the overall themes that the piece is exploring and expressing).

On a higher level, the sound serves to deepen the meaning the audience derives from each moment and of the overall story of the play (or, if you will, the program). That meaning may take the form of surprise, excitement, fascination, or the appreciation of beauty.

The highest level of purpose that sound design possesses is to create art. Sound design is an art. The true nature of the sound designer is artistic. If you are to be a sound designer, assume the role of an artist immediately. It is one of your own highest purposes.

Let us look at all three levels of purpose in more depth.

To Inform, not to Manipulate

Sound relays information in ways that nothing else can. It provides us with wonderful options that are unavailable to any other design element in a production. Tone of voice tells us whether a person is being sarcastic or not. Singing birds tell us that it is morning. The growl of a grizzly bear warns us of danger. The sound of air rushing through the vents in your home tells you that the air conditioner just kicked in. That is the first purpose of sound in a theatre event – to provide information to the audience about the circumstances of the story.

This strict information-only discipline stands firmly in contrast to a widely held but obfuscating approach that sound should be used to make the audience feel a certain way. The sound designer should never directly seek to manipulate the audience into a particular meaningful outcome. That objective is both manipulative and futile. Manipulation will not lead you to the truth, which is the only path an audience can take to find meaning. It is a misconception to think that sound will move an audience to, say, surprise or to feel heightened tension, sympathy, or any other emotion. You cannot make anyone feel anything. Every individual decides how they are going to respond to certain individual stimuli.

That is not to say that your work would not be emotionally meaningful. Concentrate on making it meaningful for you, and you will find that many people will derive the same meaning. Many more will derive other meanings from your work. In fact, it is a rare individual that will derive no meaning at all. There is a legendary theatre anecdote that goes something like this:

A young woman was cast to play the ingenue in a new play to appear in an off-Broadway theatre. In the first scene, she is having an awkward discussion with her father. At a certain point, she is to deliver the line, "May I have some more tea?" The line broke the tension in the scene. During previews, that line brought the house down with laughter from the audience. She began to deliver the line with more and more enthusiasm, and every night the audience faithfully exploded with laughter. However, when she delivered the line on opening night, to her surprise, the audience was quiet. No laughter at all. The next night brought the same silence. This went on for about a week, and the actress was beside herself with confusion. So she talked it over with the director. Why did it get a laugh during previews, but not once the play opened? The director looked at her and gently replied "My dear, during preview you were asking for a cup of tea, but now you're only asking for the laugh."

You see, during rehearsals and previews, the actress focused on the truthfulness of the moment, but once she started getting laughter, she became more focused on the impact the line of dialogue would have. She put the cart before the horse.

The same is true with sound and meaning. Your work will have an emotional impact when you focus on creating sound that will express the truth of the moment. That expression can be fantastic or humble, depending on the moment. However, if you focus on creating a specific emotional impact at the expense of truthfulness, then you have set yourself up for no emotional impact whatsoever.

No two audience members will have the same experience during a performance. A person's response to any specific stimuli depends completely on the meaning that person assigns to the event that caused the stimuli. No two people are alike; therefore, no two people will interpret intellectual information in the same way. It is equally true that no two people will respond in exactly the same way. Some people enjoy the sound of microphone feedback, and some people cringe at it and cover their ears. Same sound, two different reactions to the sounds due to two different assigned meanings.

Let us say that you decide you want the audience to feel ill-at-ease during a scene. You decide you want to do something to heighten the tension that the audience experiences. To do that, you add music that you may think of as scary, and therefore, you assume the audience will respond accordingly. Granted, maybe some will experience fear, maybe even most, but not everyone. Some people will smile because of the music you prescribed. Some people will find it annoying, and it might even lull some others to sleep. Our experiences and predilections teach us to assign meaning to events such as sound. It is a futile manipulation.

The Conditioning Factor and Mirror Mechanism

That said, there is something I call the conditioning factor. Most of our responses in life have become habits. Modern audiences, having witnessed storytelling daily for literally their whole lives in the form of television, movies, theatre, and literature, have been conditioned to follow where the production elements tell them to go. A dark stage. Haze. Music that is choppy, discordant, and in the lower frequency band. The sound of footsteps in the dark. Someone screams. The audience has learned that this is supposed to be scary, and they willingly comply. Indeed, they have purchased tickets for that very purpose. They want to be entertained and delighted. The audience is more than willing to cooperate with your prompts.

Additionally, humans have a mirror mechanism. There are mirror neurons in the brain that activate both when an individual performs an action and when they observe the same action performed by others. Along with actions, your mirror mechanism allows you to consciously and subconsciously interpret clues from other people such as body language, facial expressions, and vocal inflections. These clues cue you to attempt a prediction regarding what another individual is feeling.[1] An audience member's mirror neurons allow them to experience that same (predicted) feeling, to some degree, from the safety of sitting behind the fourth wall.

The conditioning factor and mirror mechanisms work together in an audience. You must understand the difference between them and the sincere emotion that your work inspires in someone if your intention is to create art. You can use both to your advantage, for sure, but you will never control how a person will respond to your art, and you will never inspire the same response in everyone.

Therefore, make not your goal be to "get someone to feel a certain thing." Make your goal to add sound or silence that simply tells the truth about the story you and your creative collaborators, such as lighting designers and costume designers, are telling. Allow the spectator to do with it what they will. Focusing on the communication of information in a compelling way will move people more readily deeply than will manipulation.

The Truth You Can Express

Let us say you are designing the soundscape of a scene. A telephone rings. This is a great opportunity to communicate an abundance of information about the particulars of the story. Get clear on those particulars first. In what year does the scene take place? A telephone in the year 1941 will sound different from a telephone in 1979, 2021, 2041, or 2241 for that matter. Let us say it takes place in the year 1970. What did a telephone sound like in 1970? Is it a home telephone or an office phone? Let us say it is a home phone. Is it in the USA? A telephone in the USA sounds different from a telephone in, say, Greece, Ireland, Russia, or China. Furthermore, a telephone of that era in New York City would sound different from a party line telephone in Tulsa. Let us say it is an apartment in London. Who owns the phone? Would the character who owns the phone have a modern phone for that era, or would that character prefer an older phone? Let us say modern. This tells us something about the character. Perhaps it is a newly renovated apartment. Perhaps the owner is an early adopter of technology.

What if, however, the character likes antiques? We can say something about the character if the telephone were older, and the telephone ring you choose would express that.

So far, here is a list of the information we chose to communicate:

- The year is 1970.
- The place is an apartment in London, England.
- The owner of the telephone has recently purchased a modern phone.

Once you have made these choices, then you can ask yourself how you might dramatize that sound to provide even more information to the audience. For example, perhaps the telephone happens during a particularly tense moment. You can try to interrupt that moment by creating the ring to sound sharp, sudden, and loud. You might heighten the tension in the scene by increasing the frequency of rings. On the other hand, if you choose, you can make the ring softer, so that the interruption will be more of a relief or cessation of tension, rather than adding to it.

Whichever you choose, let it be connected to how the character in the room perceives it. If you want to heighten tension, your goal would be to simply reflect what the character is feeling, as opposed to trying to make the audience feel more tension. On the other hand, if the character considers the telephone to be a relief of tension, your choices can reflect that response instead of trying to "make the audience feel relieved."

Like that telephone, underscore should be approached the same way. When you underscore a monologue and your goal is to reflect what the character is feeling at that moment, you stand a higher chance of the audience sympathizing with that character or situation than if you simply tell the audience what they should be feeling. We will talk more about this in subsequent chapters.

It might be counterintuitive, but your audience will experience much more meaning from your art if you simply reflect and express information in the context of the story than if you command how and what they should be feeling at any given moment.

Sound is a sensory art. Our senses are the only contact we have with each other. Your brains exist in a dark shell, and your senses are the only resources that your mind receives information from the outside world. It is how we receive and transmit information and achieve communication with other minds that exist in their own dark shell. Think about those senses. Your eyes receive light. Your ears receive sound. Your nose receives odor. The nerves in your fingers and skin feel things that exist in a solid state. Your taste perceives chemical qualities of the things you put in your mouth. It is your mind, in that dark shell, which interprets that sensory information and decides what it means. You decide on that meaning, sometimes consciously and sometimes subconsciously.

One person can find beauty in sound, depending on that person's taste. Some will find very minimalistic sound beautiful (meaningful), and some will find beauty in sound or music that is busy, lush, and dense. No two people have the same experience, and no two people are in the same place mentally at any given time. Therefore, no two people will derive the same meaning from the same moment. You cannot expect everyone in your audience to have the same experience.

The goal of your art should be to inspire meaning. The goal should say nothing about what that meaning will be. It is not your responsibility, so let that go. Surrender to the wonderful freedom of allowing others to take meaning from your art as they choose.

If someone is unmoved by your art, it does not mean that "they don't get it." It also does not mean that you did a poor job, you are irrelevant, or that it has no value. It simply means that it did not resonate with them at the time they were exposed to it. Some things resonate with us and not others, and some things seem to resonate with everyone else, but not us. Whether something you create inspires or does not inspire meaning is not connected to the value of what you created. Creation is always valuable. Meaning is ephemeral because feelings are ephemeral, but creation is past, present, and future.

Sound Design Is Art

Are you surprised to hear that sound design is an art? If so, you are not alone. Many people, including some sound designers themselves, have forgotten (if they even think about it at all). Sound design is an artistic expression, integral to storytelling. Audio technology has progressed meteorically as computer technology has advanced. You almost have to be a computer scientist to understand the tools we use today to their full potential. It takes a good deal of ongoing study to understand how and why a sound system works, and lots of practical experience to understand how to manipulate that system in an artful way.

While it is true that the technology of sound has advanced drastically, the advancement itself is nothing new. Sound has been an integral part of storytelling since mankind has been telling stories. Before there was the written word, there was the spoken word. Without sound, telling stories in the oral tradition would not be possible. The spoken word is sound.

Picture an ancient community gathering in celebration of a successful hunt. People dance to the rhythm of hand clapping and the rhythmic drumming, using only logs, rocks, and sticks. A hunter tells the story of what happened in the open field that day or in the woods. He roars magnificently, imitating the mighty prey. Perhaps he embellishes by imitating the animal's vocalizations as it fell. If it was raining during the hunt, he may mimic the sound of the rain and thunder. This was sound design in its most primitive form. Primitive by today's standards, but no less meaningful and exciting to the captivated listeners. Throughout history, theatre has always been accompanied by sound, and the technology of creating sound and sound effects has advanced as theatre itself has advanced.

All art forms have benefited from the technology of its time, not just audio. The invention of the phonograph has brought music and sound into the homes of millions of people (and has been raising the noise floor in the household ever since). The invention of the electric guitar inspired musicians to create a new form of music in the 1950s – rock and roll, which, so they say, will never die. The digital age provided a new way for artists to create art and music on their computers and tablet devices, inspiring new forms of visual and graphic arts. It also allowed writers and authors to type their words into a word processor, making editing and distribution faster and easier. The internet increased that distribution time to light speed, allowing readers to obtain and download books onto their tablets and e-readers in seconds. Poetry is shared on X, and photography is shared on Instagram. I get excited just thinking about what might come next.

Sound design is not just a professional trade; it is a craft. One that takes 10,000 hours to master. One that requires the artist to make countless decisions. What sounds to use? How to create them? What speakers and microphones to use to deliver that sound to the audience? At what loudness should the projected sound be? In what order should the sounds happen? What is the exact timing in which they occur? How much is too much? How much is too little? How much is enough?

Yet, there does not exist a universal understanding that sound design is art. The craft is often considered to be simply technical, and that opinion may be held by the majority of those working in the performing arts. These opinions inform the expectations of those who work with and hire sound designers. Consequently, sound designers are often expected to create the sounds, install the sound system, program the show control, and mix the performances during the run, and more. All of these differing tasks ameliorate the quality of the sound design, hindering the designer from focusing on the design itself. Engaged in all those roles, the sound designer cannot possibly devote enough time and energy to making the design great.

Make no mistake; sound design is an occupation that requires artistic expression. If everyone acknowledged that, then sound design would be allowed to reach its artistic potential.

Art is the individual and deliberate creation of something that is meaningful.

So, at this point, you may be asking yourself, "What, then, is art?" Philosophers, poets, painters, and scientists alike have pondered the meaning of art, and it can be intimidating. Here is a secret: your opinion on the definition of art is just as good as mine, and mine is just as good as yours. So, in the name of forming a common language between us, let us boil all of the theories and philosophies down to their most basic common component.

Art is simply the individual and deliberate creation of something that is meaningful. Nothing more, nothing less.

Art Is Individual

My father once told me that every joke that will ever be told has already been told. Take away the specifics of the story, and you can boil a joke down to common core components. What makes it funny is the unique take a person puts on it. Tone of voice, the tempo of the telling, facial expressions, etc. Tell the joke in a way only you can tell it.

As a child attending catholic school, I was told by a priest that humans cannot create anything. That true creation is to make something out of nothing. I think there is some truth to that. Here is an example: try to think of a primary color that does not exist. You cannot do it. The human mind does not have that capacity. Another example is to think of a truly unique living element, one that shares no properties with any other. Again, you cannot do it.

Human creativity is about connecting two things or ideas that already exist in a way that is new to you. Perhaps that combination results in something you have never thought of, heard of, or seen before. Rock and roll was a different style of music that used instruments in ways that were never used before in the history of mankind, but it still used instruments, and it was still music.

Another example is to think about aliens. Every science fiction movie you have ever seen has portrayed aliens with elements we were already familiar with. They might be lizard people, little green men, dog people, or shapeless goo. All things we already know about.

What makes it art is when you create it in a way only you can. When you tell a story in your own way, then it is art. Dance, painting, and music composition are arts when it is unique to you. It is your duty as an artist to do it your own way.

Art Is Deliberate

Art is intentionally art. We as sound artists have to look at a moment in a play that calls for an added sound and ask, "How does this sound effect help to tell the story of the play and/or the story of the moment?" What does the moment call for? If you choose, say, thunder, the choice to have thunder needs to be for artistic reasons and therefore deliberate. The qualities of the thunder must be deliberately chosen to tell the story of the moment. Why Thunder? How loud? How close? Crashing thunder or rolling thunder? Make the choice.

There is even room for art to be created randomly for the sake of being random. That is true enough. The randomness of the artwork does not cancel out the deliberate intention to create a work of art. Randomness is the intention in this case. It is that intention that is required for it to be a work of art.

Art is the expression of an idea that is meaningful.

Art is not necessarily an expression of truth, as they say, and it does not have to be in order to be meaningful. A work of fiction can be very meaningful.

That work of fiction might express some deeper truth, but there is a difference between meaning and truth. Consider the works of the great painter Jackson Pollock. There is no quality inherently truthful or untruthful about his masterpieces. It is their inherent beauty that makes them masterpieces, and that beauty is meaningful.

Do not conflate truth with beauty. Truth and beauty are two different types of expression and are not dependent on each other. Truth is often, if not always, beautiful, but beauty is not always truthful. They both, however, can be quite meaningful.

Another popular idea that we seem to take for granted is that art imitates life. That imitation is one way to express truth, and so it can be meaningful. There is a case that can be made that the abstract works of Piet Mondrian, Mark Rothko, or the minimalist art works of Agnes Martin are successful in creating great meaning without imitating life itself.

That meaning does not have to be experienced by everyone. It does not have to be a large group of people. In fact, if only one member of the human race finds it meaningful, then it is art. Even if that one person is the artist themself, that is why art is so subjective, and that very subjectivity is why it is so very hard to pin down exactly what art is. For our purposes, for the sake of a common understanding in this sound design theory thesis, let us adopt the above language and debate it in our forums forever, as it should be. This common language serves us; it has utility for our purposes. Namely, that art is simply the individual and deliberate creation of something that is meaningful.

A Logical Expression of Art

Often, philosophers will create a logical expression to use as a rubric against which to measure other concepts. Here is one we can use for the definition of art we are using for this book:

$$\forall i\ M(i) \wedge E(i) \rightarrow A$$

where i = the idea, M = the quality of meaning, and A = the art that is created. Any idea that is a meaningful idea and is expressed is art. As you create sounds for the play you are working on, you can come back to this expression to guide you. Ask yourself:

- Do I have a clearly defined idea?
- Is that idea meaningful to me?
- Did I express that idea in the sounds I have created or decided to use?

If you can answer yes to all three questions, then you know that your efforts were successful. The degree of meaning in the idea you have expressed is subjective and therefore not relevant to this expression. How can you measure something as subjective as the intensity of emotion felt by a person at any given time? You cannot, any more than a doctor can objectively measure a person's level of pain. They can only measure the subjective intensity of pain. As sound designers, we only have an individual's word regarding how much meaning they derived from your work, and it will vary from person to person. The above expression only tests if meaning is experienced in a binary way. Was meaning experienced? This expression only cares if the answer is yes or no to tell if it was successful or not.

Meaning Is an Inspired Emotion

The word meaning is widely used, and most of us intuitively know what it means, but few understand it. Think of it as that quality that inspires emotion in a human or group of human beings. Something is meaningful when it inspires an emotional response, sometimes intensely, and sometimes softly.

There is no thought devoid of meaning. Understanding always prompts emotion. Every thought a person achieves is linked to an emotional response. That includes thoughts presented to you by someone else. Understanding and meaning always walk together. It does not matter what that emotional response is. It can be amusement, anxiety, excitement, fear, joy, sadness, sympathy, or any sincere human emotion. It is part of being human. It is the awe you feel when viewing the beauty of the Mona Lisa or The Last Supper. It is the calmness you feel when listening to Debussy's Clair de Lune. It is the disgust and/or pity you feel toward the Phantom in The Phantom of the Opera. The romance you feel when you see Romeo and Juliet onstage. It is the horror you feel when watching a scary movie. It is the satisfaction you feel after reading Dostoevsky.

Have you ever asked yourself why emotions are called feelings? Emotions are experienced in with the brain and the body in tandem. We feel emotions physiologically, as the limbic system controls the release of hormones and other chemicals that affect your body. Positive emotions such as love, appreciation, and gratitude release neurotransmitters that calm the body down, and negative emotions such as anger and fear release chemicals that put the body in an alert state. We feel emotions based on the meaning we choose to place on events in our lives and the information we gather. We can choose how we respond to that stimulus based on our experiences, and those responses can be habitual. The meaning we take from it is the emotion inspired by it.

Art is the individual and deliberate creation of something meaningful. There is no quota for how many people find meaning in your art, and there is no requirement that the meaning a person finds in your work is the meaning you imagined they would experience. If you intentionally create something meaningful for one person, it is art.

Professionalism

A professional is one who lives and works according to their values.

Acquire Skill

The word art comes from the Latin word "artem." The Latin word for skill is "arte," especially practiced skill. The Latin word for practiced art is "arte." It is the same word. It is no surprise that Shakespeare uses the word art and skill interchangeably. An artifact is an object made by human craftsmanship. Artificial means to be made by human skill; a definition that then evolved to be understood as "an imitation."

The artist practices their craft to mastery. They commit to daily practice. They enjoy the process. They detach from the outcome. A professional seeks to become a master of their craft.

On February 7, 1964, the Beatles performed on the Ed Sullivan Show, but if you go back and watch the video, you will hardly hear the music above the screams of the audience. Seventy-three million people tuned in to watch. Forty-nine percent of Americans witnessed it. It was before cable TV, TV on demand, and video-streaming services. Forty-nine percent of Americans witnessed it live and at the same time. The Beatles went on to become one of the biggest influences of the 1960s counterculture, and 11 of their 12 recorded albums went to number one on most charts. Their music is still listened to by people all over the world even today. Yet, that is not the beginning of the story.

The Beatles started out seven years earlier. Paul and John met playing weddings and spot gigs, quickly joined by George. The band took action. They performed for anyone who would hire them. They did not wait to perform until they felt they were perfect. They obtained a manager who sent them to Germany, where they spent two years performing in seedy nightclubs, playing sets lasting more than 12 hours at a time. This forced them to develop their own style. When they were not performing, they went to see other English bands performing in Hamburg, met with them after gigs, picked up tricks, and incorporated what they learned into their own act.

It was only after seven long years and hundreds if not thousands of hours of grueling practice and performance did they "magically appear on the scene out of nowhere."

Stephen Sondheim, one of the world's most revered lyricists and composers of modern and post-modern musicals, started writing musicals while he was in boarding school. He was inspired by Oscar Hammerstein, his de facto father figure. Stephen was a hard worker, practicing and performing his art over a course of years before his big break writing the lyrics for the Broadway musical, "West Side Story."

Grandma Moses became outrageously famous for her paintings depicting rural American life, but she did not start out that way. Grandma Moses worked on a farm in Virginia most of her life and only started painting after she retired. She entered her work into the local county fair but never received any kind of recognition for it. For years, she painted. She practiced her art and entered her pieces over and over until one day, at one particular county fair, Louis Caldor, an art collector, happened to come across one of her paintings. While she continued painting, Caldor advocated for her, attempting to have her work placed into museums and galleries. She continued painting every day until she died at the age of 101.

The common denominator between these disparate master of their art is not hard to see. They only achieved real success after they mastered their craft, and they only became masters after hundreds of hours of work, work that mixed both practice and performance. None of them stopped once they had become masters, nor did they stop practicing once they achieved a bit of renown. They continued to practice, to learn new skills, to explore new and creative ways to create their art, and to expose their art to others.

Mastering your craft takes work. It takes daily practice. Practice includes both learning and performing. Practice and learning that continues for the length of one's career.

As a sound designer, I urge you to consider the following essentials to mastering your specific craft:

- Keep up with the latest technology and evaluate its usefulness to your art personally. Learn all about the physics of sound. Learn about how sound systems work on the electron level, and how to engineer those systems so you can manipulate them in an artful way. A way that only you personally can. These are some of the brushes you will use to create. These are the tools in the practice of your technique.
- Perform immediately. Design shows as often as possible. It does not matter if you design for a local high school or a community theatre when you are beginning. Just do the work and learn from what you do.
- Study human physiology and how the brain interprets sound, and think about how you can take advantage of that knowledge to create meaning in your shows.

- Study the physics of sound and its propagation. Learn what it is and its characteristics.
- Learn and practice other artistic disciplines. What lessons can you take from, say, lighting design, visual arts, or from practicing the piano? Perpetually explore how you can take those lessons and apply them to your design.
- Make connections with others in your field that may be doing things that you can learn from and make your own.
- Go to the theatre. See shows that you have not worked on for the lessons you may learn, and the beauty and meaning you might personally derive from them.
- Make your practice holistic. Bring everything you know, everything you feel, and everything you are learning. Bring your whole self. And show up to play every time you put yourself up at bat.
- Study playwriting and script analysis. Try your hand at it. Learn from the practice.

Master your craft, and once you do, continue to learn, practice, and perform. See how far you can take your art. Contribute to the conversation. Along the way, see how much meaning you can create for yourself and your audience.

The OK Plateau

In the book "Moonwalking with Einstein," Joshua Foer describes what he calls the "OK Plateau."

> *The OK Plateau is that point when we reach the autonomous stage and consciously or unconsciously say to ourselves, 'I am OK at how good I have gotten at this task,' and stop paying attention to our improvement. We all reach OK Plateaus in almost everything we do. We learn to drive when we're teenagers, and at first we improve rapidly, but eventually we are no longer a threat to old ladies crossing the street, and we stop getting appreciably better.*[2]

This plateau is natural and quite common, and therefore, as a professional seeking mastery of their art, you must always be aware of it. The moment you sit back on your heels, dwell in your past accomplishments, and silently say "that's good enough" is the moment you stop making art. As humans, we are biologically rewarded when we grow. We have evolved to feel good when we are making progress at something and to feel bad when we stagnate. As a professional artist, you must always practice, always experiment, always push yourself to get better at your skills, and to learn new ones. It is the happiest task we will ever know.

Growing is a bit of a risky business. Your audience wants new versions of what you have done before. Your reputation is at stake, after all. Money, too, is at stake. Yet you will always be capable of more. Humans change, learn, and evolve. If you make your goal to show up for your art, it, too, will evolve. That evolution is good for you, and good for the world that will find meaning in your evolving body of art.

The Second Part of Professionalism: Values

Professionalism is simply living in accordance with your values and bringing those values into your work. Yes, it is about dressing up on a job site, but only if you value looking good. There are reasons to look good. You want people to regard you with credibility. Looking nice gives the impression that you have your act together and can be trusted to do your job. It is because one of your values is to look "professional." Over the years, the suit and tie have been replaced by a t-shirt and jeans, but it is only the fashion that has changed. The idea of dressing appropriately has ot changed.

Yet, dressing appropriately and comporting yourself in a courteous and helpful way is not what I am talking about. Let us look at what it means to have professionalism as an artist by looking at the values and artist must hold.

- An artist values skill in their craft. They practice routinely to get better at technique. They read. They talk to people. They do not stop learning once their career is launched. They work their whole career to become a master of their craft.
- An artist values free expression. How can they express meaning in their own life if they are compelled to keep limits? They use free speech wisely.

The Collaboratory Choices Are Yours

As discussed, the sound designer is responsible for everything the audience hears and how they hear it. At the same time, theatre itself is collaborative. Theatre is created by a group of artists practicing different disciplines who get together and work together to build the story according to the Director's vision. The team shares their work and shares ideas. Ideas and possible choices get floated and bounced around. Take the director's ideas and run with them. For that matter, take good ideas from wherever they come. However, always know that the final decision about whether to use a good idea or not rests with you, but only if the director agrees with you. The director should always have the last word, even if you do not agree with them. One of the director's responsibilities is to lead the team to work together. A good director will bring out the best of each member of that team while at the same time ensuring that everyone on the team is doing the same show, guiding the team to realize the director's vision for the show. If you and the director aggressively disagree on something, in the end, you honor the director's decision.

When you work with a group of people, there will be those (who are not the director) who will tell you what you should do with sound. They will happily give you notes on how to tell your story. If their ideas are good ones, take them. If you think they are not so good, you are under no obligation to adopt them. Consider them suggestions. Always remember that it is your duty to make choices of your own. Take a good idea from wherever it comes, but you are not obligated to take any bad idea.

At the same time, be careful not to make decisions that are not yours to make. Do not tell the lighting designer how to light a moment, but you can talk with them and discuss your intentions and ask that person to work with you.

The same goes for your mixing engineer and all other members of the sound crew. There are decisions you must make to realize your design, and there are decisions you should leave in the hands of the engineer. Hopefully, you have mastered the art of live mixing with years of experience under your belt so that you can bring that skill into your designs, but you are not mixing this one. Let the engineer program the console their own way. Tune the room with them. Allow your engineer to lead the other members of the sound crew the way they see fit, according to their style and personality. Certainly, you oversee all aspects of the sound department, and if you feel strongly about a certain decision and how it affects your intentions on how the show should sound, it is your duty to make the engineer aware of it and give suggestions on how to mix it. This might be a certain EQ curve, an expander, how to pan instruments, even a bit of programming, or an entire programming approach.

Knowing which decisions are yours to make, and which are collaborations, is all part of your art. They should all serve the goal. Knowing which decisions to take and which to leave to your team has only one measure: Which would be best for the show?

Exercises for Chapter 2: Foundational Concepts

Exercise 1: Journaling Challenge

Journaling helps us to explore ideas and how to express them. It helps us understand how we feel about the world around us. It is a good daily habit to adopt. I challenge you to write 500 words every day. It is enough to force you to go deeper than your first impressions and to think about your next thoughts. It is not so much that it will overwhelm you. I personally do write 750 words daily, seven days a week.

When you journal, write without stopping or editing. Keep going. If you write something that you do not think is true, note that you do not think it was true. Revise it, then again, as many times as you have to, until you feel that it is what you want to say. Try not to write what you think you should be feeling or thinking. Try to write how you *actually* feel about it and what you actually think about it without judging yourself.

You do not have to share it with anyone. Write with the belief that no one will ever see it. Put a password on your journal. I like to believe that no one will ever read my journal. This allows me the freedom to explore what I think is true without worrying about being judged by others.

Exercise 2: Sound Design Choice Analysis

Objective: Understand the impact of authentic, holistic choices in sound design as outlined in this chapter.

Instructions:

1. Select a scene from a play with a prominent sound element (e.g., a monologue or a dramatic moment).
2. Design a sound cue for this scene, considering your personal experiences and tastes.
3. Write an analysis explaining how your choices reflect your unique aesthetic and how they serve the story's truth, avoiding manipulation.

Exercise 3: Journaling Exercise: Defining Personal Artistry

Description: Begin a regimen of self-reflection.

Instructions:

Journal 500 words on what "art" means to you as a sound designer. Describe how your personal experiences, values, and creative instincts shape your approach to creating meaningful sound designs. Discuss how you can avoid the "OK Plateau" by committing to lifelong practice. Reflect on your identity as a sound design artist. Describe your artistic values. Journal about what first attracted you to sound, and whatever you feel is important. Add this exercise to your regular journaling habit.

Exercise 4: Professional Values Mapping

Description: Develop a personal code of professionalism based on this chapter's discussion of values and skills.

Instructions:

Create a visual or written "values map" listing five personal values that will guide your work. This, too, is a topic you should include in your daily journaling. You must feel free to add to this list as you grow and learn more about yourself. Anything on the list can be revised.

3

Intentions and Choices

People often conflate the concepts of artistic intention and artistic choice. The terms are often considered to be the same thing. They get tossed around in production meetings with a vague understanding of the difference, but the nuances of what those terms really mean remain unclear.

Intentions deal with the future. Choices exist in the past and also in the present moment. Intentions are what you plan to do. Choices are how you do it. Intentions and choices are your keys to expressing your meaningful ideas.

In this chapter, we will explore the concepts of intentions and choices, and how they are shaped by context and constraints. We will explore a bit of theory and then apply that theory in practical applications.

In the last chapter, we explored the following logical expression:

$$\forall i\ M(i) \wedge E(i) \rightarrow A,$$

which declares that any idea that is meaningful and is expressed is art. As sound designers, the meaning derived from our art form is derived from the choices we make in the act of expressing our meaningful ideas. The intentions we set and the choices we make are the building blocks of our final product. There is no meaningful expression without intentions or choices. A painter must choose their colors, a dancer must choose their movement, a musician must choose their notes, and a sound designer must choose their sound. Unless you set deliberate intentions and make deliberate choices, any

DOI: 10.4324/9781003519393-5

meaning that is derived from what you produce will be accidental and that is not from you, and you have not expressed anything.

Freedom of Expression

If art is a meaningful expression of a meaningful idea, then it stands to reason that sound designers, along with all artists, should hold the concept of freedom of individual expression among our highest ideals. That means that an artist must value the freedom of expressing themselves, along with the freedom of all others to express themselves in a way of their choosing. To share your idea in your own way is always for the good, as long as you realize that you cannot force anyone to agree with you or be moved by the idea. To talk and to listen are what freedom of expression is all about. We all grow when meaningful ideas are shared. When we talk and listen, we enlighten ourselves to a new perspective. That perspective allows us to derive even more meaning from life, and meaning is why we are here. This new perspective motivates us to form and pursue our intentions via the choices we are free to make. Meaning is derived from communicating in our own way. Meaning is derived from the choices we make in our art.

Since art is the expression of meaning, or a meaningful idea, and meaning is an inspired emotion, and emotion is a choice based on input and beliefs, then your art and your character are intertwined and cannot be separated. By character, I mean the set of values and beliefs you hold. You cannot separate your art from your character, since you need both an expression and an idea to express, and the artist must find it meaningful. The artist creates art because he or she has the desire to share that meaningful idea with others or with themselves. They do it in the hope, and not the expectation, that someone else may be moved by it.

The intentions that you select are based on the values, beliefs, and thoughts that you find to be true. You cannot get away from it. Even if you are expressing someone else's ideas and even if you disagree with those ideas, you are selecting to express them for a reason. These reasons are your own that are, again, based on your values, beliefs, ideas, etc. A lawyer, for example, argues for the client, even if he disagrees with that client. Why? Because the lawyer believes in a higher ideal, probably that advocating for someone less skilled or knowledgeable about the law is a mission aligned with the lawyer's values and beliefs. Another reason an artist may choose to express other opposing ideas is to be ironic. Irony, by definition, is the expression of an idea by using language that normally means the opposite. In this case, the artist is actually expressing an idea that he or she finds meaningful, and the use of irony is a choice.

Give yourself permission to boldly express your own ideas in your own way. Do it for the value of the expression itself. Do it not to be relevant, admired, or for any other reason but to simply honor the idea that you hold true. Some people will hate it, but not everyone. Some people will love it, but not everyone, and some people will be ambivalent, but not everyone. Let your intentions be your intentions and your choices be your choices. In the process of writing this book, I remind myself, "It may go down in flames, but at least it will be because I stated what I believe as truth, and not because I'm simply creating content that people will reward me for." I don't want to create clickbait.

Intentions

Intentions are broad. Your design is the plan of sounds that you would like the audience to hear. That sound is the expression of an idea, which, as we have said, can be either a feeling or a concept involving a feeling. Examples of feelings can include regret, contentment, loneliness, love, etc. Feelings experienced in circumstantial contexts can include things like determination in the face of overwhelming odds or the appreciation of a mentor. The awe in the appreciation of beauty. Obviously, there are countless examples. If you have made the decision to express an idea like this, then you have set an intention. In contrast, how you are going to express that idea is a choice.

Intentions are what you plan to express and therefore exist in your future. You can attempt to set an intention for this moment, but by the time that intention is manifested, that moment is gone, and another has replaced it. Additionally, there is the practicality of creating a sound design. We sound designers plan in advance so that we can attach our choices to the program. An intention comes before the choice, and the choice comes before the performance of that choice.

Just for fun, you may ask, "What about improv?" Improv employs the same process of intention, choices, and performance of sound. It just happens at lightning speed.

Intentions are bound to the beats of the script. Each moment must have its own intention, since each moment is different. Every moment is unique, and therefore, each moment is an opportunity to express an idea, even if that expression is silence. Silence is not the absence of expression; silence is a choice, and therefore, every bit an expression of an idea as sound is. After all, the play itself is a procession of individual moments. Each of those moments is enveloped in sound or silence. These moments stack, one on top of another, in harmony with the journey that the characters take. The journey that, hopefully, ultimately, and unavoidably changes the characters. Each moment stands on the shoulders of the moments that came before.

Intentions take a broad view and require an awareness of the events of a moment and the meaning those events inspire.

Choices

While intentions are broad and can sometimes be sweeping in scope, choices are granular. A choice is how you decide to express your intention. It is whether to create sound for a moment or not. A choice is the sound you create, and choices usually include timing, tempo, timbre, and all the other qualities of a specific sound cue. Is it music? Is it recorded? Is it live? Is it loud and low, like the growl of a trombone, or is it soft and treble, like the soothing chirp of a cardinal? What precedes the sound or silence? What is the relationship between your choice and what precedes its delivery to your audience? How long does it last? Does it sneak into the audience's perception, or does it break through the door like a rhinoceros? In the end, is it meaningful to you, and does it inspire the emotion in you that you intended?

Choices exist in the present moment and in the past. They must be deliberate and therefore your own. Any choice left up to chance is not your own, even if it accidentally inspires meaning in you or in others.

Choices stack. In sound design, these ideas are nested and related to a specific moment. Once the moment is gone, so too is the sound. Since sound is an expression, you can say that once the moment is gone, so too is the expression of that idea. When the next moment brings the next expression, the two moments combine to create a more complex idea.

A choice, then, is the sound or silence we choose to express a meaningful idea in a specific moment of the production. An intention is what we plan to express in the future. The choice is the expression. An intention is the idea.

The Freedom of Constraints

Constraints create the opportunity for unlimited creativity. They set the rules under which your choices are made. Choices are idealistic and practical at the same time. We aim to make the best choice given the rules we decide to adhere to, and they can be big like a symphony, intimate like a quartet, fantastic like a thunderstorm, or subtle, like the whir of a hummingbird's wings.

We attempt to create the most meaningful sound, and every sound you make comes with constraints. However, as with the Haiku, constraints are a crucial ingredient to manifesting your choices into reality.

All constraints exist as either one of two types or both.

External constraints are those that come from circumstances or people. Theatre productions exist in a practical world and come with practical constraints such as budget, supplies, time (deadlines), and other limited resources at your disposal. These are constraints that are simply not in your control. Other constraints in this category include members of the creative team, the director's vision for the show, the story they want to tell, themes that the director wishes to explore, the venue in which the show will be performed, etc.

Internal constraints are self-imposed. These are the constraints you place on your work as you make your design choices. Have you chosen to create sound effects that adhere to the time period or era in which the show is set, or is it your choice that some or all of the sound cues will be anachronistic? Do you choose to create music to represent sound cues? Perhaps you choose not to use any recorded sound, and that all of the sound cues are effects created by people backstage or even onstage. Perhaps you decide you would like to have musicians play all of the music you have written for transitions from scene to scene? Other internal constraints might be self-imposed deadlines you create as you create your project plan to meet the external deadline. Perhaps you decide to have a certain percentage of the work be completed each week so that you are done by the time that the external deadline arrives. Personally, I like to have a first draft of all the sound and music cues complete by the date of the first rehearsal so that the director, actors, and stage managers can rehearse with them. This gives the show a chance to evolve during the rehearsal process. The actors can work with the cues. I can change the cues based on what the actors and director discover. Other designers find that the first draft is helpful as well, so they can understand what direction the sound elements of the show are going in. This self-imposed deadline enables a more collaborative environment and sets all of us up for a more productive tech process.

Constraints are wonderfully freeing because they guide us. Think of them not as limitations, but as rules of the game you are playing at the moment.

Have you ever encountered a situation where a choice is being discussed, and someone in the discussion abruptly declares that it simply cannot be done? They cite a lack of resources like money, time, or personnel. When I encounter that situation, I reject the notion that it cannot be done and instead replace that declaration with the question "How can it be done, given the constraints at hand?" It is a magic question. One that allows all options to be considered. Even ridiculous options are on the table, to be accepted or rejected. Ridiculous options can be quite fruitful because they often lead to options that are obtainable that you might not have thought of. In fact, many sound designers, including myself, play a game called "What's the worst sound I can introduce into this moment?" Usually identifying the worst sound leads to an understanding of why it is the worst sound (inappropriate, etc.) and that understanding usually leads to an idea of what might be right.

Then, there is the wonderfully freeing constraint of identity. Here is the idea: The law of non-contradiction says that something cannot both be and not be at the same time. Something cannot both exist and not exist at the same time. That implies that there is no such thing as a thing that is other than itself. A thing, big or small, is exactly the size of itself. A thing is exactly its own size; otherwise, it would be something else. A thing cannot be both itself and its opposite. Think about it; no one can create something that does not have an identity constraint. If you create a painting, then you have created a visual work of art, to be seen but not heard. If you create a piece of music, then you have created a thing that is heard but not seen. A painting cannot be a video, and a video cannot be a painting. A painting is static. A video has some movement and evolution, even if it is constrained to only that movement and evolution. A video requires time, and a painting does not. If it were possible to stop time, the video would cease to exist, but the painting would not. (Although if you were to stop time, then you would also stop the movement of photons bouncing off the painting and into your retinas, but the painting would still exist, nonetheless. A tree that falls in the forest and nobody hears it is still a tree, and it still falls.)

A thing cannot be both a thing and a different thing at the same time. This is an identity constraint. Even the great creative wisdom of the universe, which created things out of no things, adhered to that rule.

All things have specific and unique characteristics. You can think of those characteristics that make a thing itself and nothing else as constraints. It will serve us in our art. An orange is a thing that is a fruit, that begins green, ripens into the color orange, and if left alone, eventually shrivels up and dies. Those are the constraints under which an orange exists. Without those constraints, it would not be an orange. An orange is a thing. A pear is something else. There are characteristics unique to a pear, and there are characteristics unique to an orange, and then there are characteristics that are shared by both pears and oranges.

Sometimes the uniqueness of that thing is the combination of what it is made of. The recipe. The proportions of each ingredient in relation to all the other ingredients. An orange has pulp, vitamin C, absorbs, and reflects the right combination of photons that make it look orange. The same can be said of a pear. However, as we know, an orange and a pear are not the same thing. The difference is realized by the ingredients that make them up, and all the other properties they possess, such as atomic signature, texture, or juiciness.

Another example is the music you play under a line of dialogue in a play. It has length, it has volume, it has notes, and the sequence in which they are played, along with all the other characteristics a sound cue has. Change one note, change the tempo slightly, or change the length, and you no longer have the same piece of music. Here is the point: it is the specific combinations of constraints about a thing that allows it to exist and to be the unique thing that it is and nothing else.

There is an old chestnut that says: If you try to write about everything, you end up writing about nothing. You must make choices regarding what to write about. Every choice you make is a constraint. In that way, constraints help us to be more productive in our productivity.

You can make a choice without knowing what jewel will sprout from it, by the way. The choices you make, each and every one, is another ingredient in the thing. Every choice you make is another characteristic. Every choice leads to another choice, and then another until you feel what you have created is meaningful.

Context

Sound in theatre is inherently tied to the context in which it is presented and uniquely affects the meaning of that cue. Context includes countless circumstances, like dialogue and action that happens just before and during the sound cue, and even expectations of what is going to happen next. Along with dialogue and action, context can include other sound cues that have come before this one, the way an actor is delivering their lines, and even the style of the play that the director is going for. Again, there are countless contextual data that might apply.

Context also includes the function of the passage of time while the sound cue itself is playing. This, too, relates to our definition of art as an expression of a meaningful idea. In theatre sound, that expression can only exist in one finite moment of time. That is because sound can only exist in the confines of time in motion. (If you stop time, sound does not exist.) Still, time passes over the execution of the sound cue. A sound cue is a chain of moments from beginning to end. The beginning of that cue in motion is a moment. Once we are in the sound cue, the beginning moment is replaced by the next, and then the next, until finally we experience the moment that the cue comes to an end. The sound you hear in the first second of a sound cue is not the sound you hear in any moment of the rest of the sound cue. They may even be identical, but their juxtaposition makes them unique from each other. Any given moment's position in the series of moments that make up the sound cue is context.

As a simple example, consider a sound cue that is simply a 1 kHz tone. How long does that tone last? Is it half a second (sounds like a beep) or a minute? Since the word "tone" suggests something that lasts longer than a beep, for this example, let us make it a 1 kHz tone that lasts for a minute – 60 seconds. If you are sitting in the audience, during the first moment of that cue, the tone is going to hit your ears. Perhaps in the next moment of the cue, you are going to hear the tone added to whatever bounces off the nearest wall. In the next moment, you are going to hear the tone added to whatever bounces off all of the walls, the ceiling, the floor, and anything around you. All that energy is going to start combining, so that in the next moment of the cue, the quality and maybe even the timbre of the cue is going to change. In the first moments of the tone, you may feel surprised, but after 30 seconds of it, you may feel bored or irritated. Finally, the cue is over. (By the way, the same concept applies if you are wearing headphones. You can never escape the effects of your surroundings, even with headphones.) It is important to note that while you will perceive that the sound is different from moment to moment, you will likely consider it unchanging. The Gestalt effect enables us to perceive the cue as one complete whole.

You can consider a 1 kHz tone as the expression of an idea. Context always informs the idea.

- If the scene is set in a recording studio, perhaps the idea of that tone is that normalizing the gain structure of the recording can be intimidating to the talent.
- If in the scene a robot is speaking, perhaps the idea of that tone is a malfunction.
- If the scene is set in an intimidating environment, such as a mob of people surrounding you, perhaps the idea is that the tone becomes a metaphor for the feeling of anxiety the character is experiencing.

Meaning and Context

When you take any moment of that sound cue and compare it to any other moment of the sound cue, they may be identical when observed out of context. When you measure them through the lens of our artistic expression, though, no two moments are the same. That is because art does not exist in the expression of any old idea. It is an expression of a *meaningful* idea. The meaning attached to the sound is different from moment to moment.

A sound cue's context always relates to other meaningful ideas that have been expressed prior to this sound cue. Like the moments of a single sound cue, meaningful ideas stack, and that relationship between ideas can sum and become a bigger idea. Consider a scene that begins with two brothers sitting on a porch smoking cigars and drinking scotch. They talk about some great accomplishment they made that day. Your intention for that moment is the feeling of brotherly love, and your choice might be the lulling nature of evening crickets and cardinals singing. Then a beautiful woman enters wearing a sundress and congratulates one of the brothers on the accomplishment, but not the other. Your intention for that moment is the feeling of jealousy, and your choice would be that the sound of the crickets becomes more intense and the chirping becomes angry. Put the two intentions together, and you can end up with a feeling of sibling rivalry. After she leaves, the second brother becomes argumentative with the first, and the conflict escalates. What would your choice be then? Since we have already created the feeling of sibling rivalry, we can underscore with some music, and that music is justified by the context of the previous moments.

Context can be scenic or setting-related, such as the sound of rain in the memory section of "Three Days of Rain" by Richard Greenberg.[1] That same sound effect will take on an entirely different meaning in the opening scene of *The Tempest* by William Shakespeare. Context can be relational. That rain sound effect can be serene, but adding the sound of thunder, it takes on a different meaning. Context can even have spatial considerations – is the rain in the east or the west? It tells a different story if the rain is moving toward the characters than if it is moving away.

Let us take a look at some examples of the context of an expressed idea as it relates to the craft of sound design.

The 60-Second 1-kHz Tone

The situation, again, is that our hero is surrounded by a shouting mob, and the sound of the mob morphs into a 1-kHz tone. The sound of the mob consists of shouting, banging sticks together, stamping feet on the ground, etc. Perhaps a fight or two breaks out among some of them. Those sounds are part of the context in which the 1-kHz tone is presented to the audience. The mob itself is part of the context – several people crowded together. The physical activity of the mob in close quarters is context. Anything we know, or do not know, about the characters that make up the mob is context. Our hero's character is context, and so is her history. Perhaps the hero has been trained to deal with mobs. That would add context to the sound, and the meaning derived from the tone might therefore be one of calm under pressure. On the other hand, if we know that as a child our hero found herself in the middle of a mob that trampled and killed her parents in front of her, then the meaning derived from that tone may be anxiety and fear. Meaning is directly related to the context in which it is expressed.

Setting Your Intentions and Making Your Choices

Everything in this chapter up to this point has been theory. The best way to put this theory into action and to start setting intentions and making choices is to start by asking yourself some questions and answering them as truthfully as you can. The thing about art is that there are no right or wrong answers. It is okay to give yourself time to think about these questions. We can only answer these questions the best that we can at this time, and we are allowed to change our answers as many times as we want after some thought.

Setting Intentions

When you set intentions for your sound design for a given program, set intentions for the entire arc of the play and for each specific moment. Ask yourself the following questions:

- What are the circumstances of the specific moment? What is happening?
- What information might I want to express by introducing sound into the moment?
- What ideas would I like to express? (This is where you intersect information with meaning.)
- Would silence, or at least opting not to introduce a sound cue in this moment, serve my intention?
- How does this intention fit in with other intentions for this production?
- How does this intention fit in with the intentions of those of other artists, such as the director, actors, musical director, lighting designer, or anyone else who is contributing to this moment?

Choosing Constraints for Your Choices

Identify your constraints before you make your choices. It is easier to create something when you define the rules. Rules help the creation take shape. Your job is to define the rules, and it must be deliberate. Be aware of the difference between when you are inventing the rules and when you are accepting the rules that others may have taught you. You can adopt, reject, and embrace rules that you think will serve you. Other constraints will be out of your control, such as an actor who cannot sing, or your preferred placement of a speaker is impossible. Give yourself permission to define the constraints that are in your control, to accept the constraints that are not, or some combination. Make the choices consciously. Be aware that you have made the choice. It is much more meaningful if you do. The meaning is in the choice. The meaning is derived from the choice, and you can do anything you want to. In fact, it is your duty to do so. Ask yourself the following questions:

- In your wildest fantasy, what sound would you put into this moment? How would it be delivered to the audience?
- What are the challenges to achieving that fantasy?
- Make an exhaustive list of rules that you must operate under that are not of your choosing.
- Identify rules that you would like to operate within.
- Ask yourself, "How can I achieve that fantasy within the rules I am subjected to?"

Making Choices

Choices can only be made for the moment being considered. All moments must be considered. Here are questions to help you make choices:

- Which sound best delivers information in concert with the action, dialogue, setting, location, period, character attributes, or other areas about which you would like to provide information?
- What sound best serves my intention for the moment?
- Does silence better serve the moment?
- Where does the sound come from? Where is it imaged from?
- How long should the sound cue last? From which line of dialogue or action should the sound start, and from which line of dialogue or action should the sound stop?
- How loud is the sound?
- Does the sound have a sense of movement?
- How intense or soft would this sound feel?
- Keeping in mind that everything the audience hears should be a symphony of music and sound, should the sound be more recognizably on the side of traditional music or on the side of sound effects?

Between Stimulus and Response, We Have the Freedom to Choose

In his masterpiece book *Man's Search for Meaning,* Viktor Frankl writes the following:

> *We who lived in concentration camps can remember the men who walked through the huts comforting others, giving away their last piece of bread. They may have been few in number, but they offer sufficient proof that everything can*

be taken from a man but one thing: the last of the human freedoms – to choose one's attitude in any given set of circumstances, to choose one's own way.[2]

Man's Search for Meaning" by Viktor E. Frankl
Copyright ©1959, 1962, 1984, 1992, 2006 by Viktor E. Frankl
Reprinted by permission of Beacon Press, Boston

Stephen Covey muses the following about the above passage in *The 7 Habits of Highly Effective People*:

In the midst of the most degrading circumstances imaginable, Frankl used the human endowment of self-awareness to discover a fundamental principle about the nature of man: Between stimulus and response, man has the freedom to choose. Within the freedom to choose are those endowments that make us uniquely human. In addition to self-awareness, we have imagination – the ability to create in our minds beyond our present reality. We have conscience – a deep inner awareness of right and wrong, of the principles that govern our behavior, and a sense of the degree to which our thoughts and actions are in harmony with them. And we have independent will – the ability to act based on our self-awareness, free of all other influences.[3]

Used with permission of FranklinCovey Co.

Between stimulus and response, we have the freedom to choose.

For every moment that you live, you have a new opportunity to make a choice. Covey's stimulus can come in many forms. A stimulus means something that causes a response or an action. It can be an event that happens to you from an external source, such as a car accident, a mosquito bite, the kindness of a stranger, or a birthday gift from your one true love. On the other hand, stimulus can come from yourself, such as a specific belief about the world, a thought on your mind at a specific moment, or a feeling you are experiencing.

Your response might be a knee-jerk reaction. It may be thoughtless, but it does not have to be. You have a space to choose your response.

Between stimulus and response, we have the freedom to choose.

You have the freedom and ability to choose your response, your beliefs, and even your emotions. In the same way, you have the freedom to choose the constraints within which your art will be created. You have the freedom, and even the duty, to choose its structure, and you have the freedom to choose whether to take the structure from norms or to make up the structure whole cloth.

Meaning is not derived from the structure. Meaning is derived from the act of choosing the structure. Meaning comes from making your choice. With no deliberate choice, there is no meaning. When you tacitly accept artistic constraints that others have defined as implicit, you are robbing your work (and your audience) of all the meaning that will exist when you actively choose.

Exercises for Chapter 3

Exercise 1: Setting Intentions for a Scene

Objective: Practice defining intentions for sound design based on a specific moment in a play.

Instructions:

1. Choose a short scene from a play.
2. Read the scene carefully and identify a key moment where sound could enhance the emotional or conceptual impact.
3. Answer the following questions from the chapter's "Setting Intentions" section in writing:
   a. What are the circumstances of this moment?
   b. What information or idea do you want to express through sound?
   c. What emotion or concept (e.g., loneliness, awe, tension) are you aiming to evoke?
   d. Would silence better serve this moment? Why or why not?
4. Write a reflection explaining your intention for the sound in this moment and how it aligns with the play's broader narrative or themes.
5. Share your answers with a partner or small group and discuss how your intentions differ or align.

Exercise 2: Exploring Choices Through Sound Selection

Objective: Experiment with making deliberate choices to express an intention through sound.

Instructions:

1. In the scene from the last exercise, select a moment for which you have set an intention.
2. Create three different sound cues to express this intention. For each cue, vary at least two elements (e.g., timbre, tempo, volume, or instrumentation). For example:
   a. Cue 1: Soft, slow strings fading in.
   b. Cue 2: A low, pulsing drone with subtle high-pitched tones.
   c. Cue 3: A rhythmic, percussive sound with a gradual crescendo.
3. In a few sentences, explain how each cue reflects your intention, addressing the following:
   a. Why you chose specific sounds or qualities (e.g., loud vs. soft, music vs. sound effect).
   b. How the cue interacts with the imagined dialogue or action.
   c. Whether silence was considered and why you chose sound instead.

Exercise 3: Constraints as Creative Tools

Objective: Embrace constraints to spark innovative sound design solutions.

Instructions:

1. Imagine you are designing sound for the scene you selected in the first exercise. Apply the following constraints to your scene:
    a. Budget: $50 for sound effects or music.
    b. Equipment: Only a single speaker is available for playback.
    c. Time: You have one day to create the cue.
    d. Director's request: The sound must feel "otherworldly."
2. Apply all you have decided about the above scene based on your answers in the prior exercises, add the external constraints above, and then create a sound cue that expresses your intention for the scene or moment.
3. Answer the following questions:
    a. What is your "wildest fantasy" sound for this moment?
    b. What are the challenges to achieving that fantasy, given the constraints?
    c. What rules (imposed or self-defined) are you operating within?

Exercise 4: Contextual Sound Design Analysis

Objective: Analyze how context shapes the meaning of a sound cue in a theatrical production.

Instructions:

1. Watch a recorded performance of a play or listen to a sound designer's cue list for a play. Focus on one specific sound cue.
2. Write an analysis addressing the following:
    a. What is the context of the cue?
    b. What idea or emotion does the cue seem to express, and how does the context shape this meaning?
    c. How do specific choices (e.g., timing, volume, timbre) enhance the cue's effectiveness in its context?
    d. If you were the sound designer, would you have made different choices (e.g., silence, a different sound)? Why?

4

From Sound to Meaning

To become proficient, you will have to learn to use the tools of the craft, such as loudspeaker systems, recording equipment, and all the other gear we can utilize to create and deliver our sound and music. Furthermore, you will need to be able to decide when, how, and why to use these tools. If you are committed to becoming a master of your craft, you must understand the most important mechanism you have: Your own brain. You choose your emotions before you experience them. Your brain chooses your emotion based on what it thinks is the most likely cause of the sensory input it receives. Understanding this process helps your work become more efficient, more creative, and exponentially more meaningful.

Your brain lives in a protected isolation tank that is your head. It is quiet, dark, and comfortable. It is surrounded by a thin membrane called the pia mater and then a scaffolding-like membrane called the arachnoid mater filled with cerebrospinal fluid. The next layer is the dura mater, a fibrous layer that lines the inner surface of the skull, providing structural support. Outside of that is the skull, tissue, flesh, and hair. (At some point in most people's lives, they have had hair on their heads.) The brain knows almost everything that is happening inside your body. For example, it knows what your heart is doing, what your lungs are doing, and what your gut is doing. Conversely, it knows nothing about what is outside your body, except through the signals it receives from your eyes, ears, nose, tongue, and the things that you touch. It does not know the cause of these signals; it only receives them. Your brain makes guesses because it wants to choose the best way to react. The brain's

DOI: 10.4324/9781003519393-6

goal is to stay alive and healthy long enough for your offspring to raise their own offspring. That is your brain, not necessarily your mind.

In this chapter, we explore how our brains create meaning (emotions) based on sensory stimuli from the outside world.

- A brief overview of how the normal human brain receives sound transmissions.
- A brief overview of how the normal human brain interprets and identifies the signals they receive.
- A brief overview of what is involved in the recognition of the sound signals our brain receives.
- Finally, we will summarize the current science on how we choose the meaning in the sound signals we receive and why.

As we cover this material, keep a few questions in your mind: How can I use this knowledge to become a better sound designer? How can this information be used and manipulated to enhance my art discipline? How can this knowledge inspire me to come up with ways to create meaning in my own unique way?

At the end of the chapter, a list of recommended reading is provided. Should you want to take a deeper dive into how humans choose emotions based on our sensory services?

Human Sound Reception

Imagine a chair in the center of an empty stage, and a piano upstage and to the right. You sit in the audience waiting for the recital to start. The pianist walks on and sits at the piano. A cellist walks out with her instrument and sits. The pianist plays an A, and the cellist prepares to tune. She scrapes her bow over the fattest shiny metal string. This friction causes the A string to vibrate, which causes an identical disturbance in the air molecules in every direction. Those air molecules collide with the molecules closest to it, and they collide with the next, and they with the next, until the energy travels all around the theatre, losing a bit of energy with every collision. Some of the air molecules cause the cello's body to resonate, amplifying and enriching the sound we hear. If you could see the disturbance, it would look like a sphere, growing in diameter in every direction, until finally all the energy is exhausted from the originating scrape of the bow and all the air molecules return to a state of rest. The pianist and the cellist then move to the D note, and the process repeats.

To understand how our brains receive sound signal transmissions, we must first have a basic understanding of what sound is.

Your brain receives sound signals like a sonar specialist in a submarine, listening to pings bouncing off the things around it. Is it a whale or an enemy sub? Is it a hill or a crevice on the ocean floor? The operator learns how to tell the difference, so they know how to react or ignore it. So do we with sound.

Sound is a vibration of molecules through a medium such as air that travels in mechanical sine waves at frequencies audible through human hearing mechanisms. Think of sound as an oscillation of pressure caused by the projectiles that are the air molecules. Energy propels one molecule into the next molecule and then backs the other way to hit the molecule behind it. It goes back and forth like that until enough energy is lost, and the molecule again comes to rest. Positive pressure is compression, and negative pressure is rarefaction.

Imagine, if you will, an empty room. The door to the room is closed, but the one window in the room is open. Imagine the door is a swinging door that moves in both directions. From the top of the open window hangs a drawn blind. When you push the door open to enter the room, the blind pushes backward toward the window. The door pushes the air forward and applies that positive air pressure to the blind, that's, compression. On the other hand, if you pull the door backward to enter the room, the blind pushes forward into the room in response to the negative pressure applied to the blind, known as rarefaction. If you were to repeatedly push the door and then pull the door, the blind would oscillate back and forth due to the repeating compression and rarefaction you create in the room.

The rate at which the door pushes forward and then backward is its frequency. One full oscillation from the middle, then forward, then backward, then back to the middle per second has a higher frequency than one full oscillation per minute. A sound's frequency is the number of oscillations of air compression and rarefaction completed in one second. One oscillation, or cycle, per second is measured as 1 Hertz (Hz). The frequency of the sound determines what we perceive as pitch. For example, a 220-Hz sound wave corresponds with the musical note on the open A string of a cello. Humans have a range of frequencies that they can perceive acoustically. Normal, healthy human ears can hear a range from *circa* 20 Hz to 20 kHz (20,000 Hz).

A sound's amplitude, acoustically speaking, is the measure of the energy that is injected into an air molecule. Not to be confused with frequency, amplitude is not the speed at which the molecule travels, but the force with which it bangs into other molecules. More amplitude causes one molecule to push the next further forward. Then comes the equal and opposite reaction and then further backward. Amplitude does not affect frequency. The speed at which the molecule makes its journey of compression to rarefaction and back to its starting point remains unchanged. Think of it as how much force you use when you push and pull that door, which will result in how wide the door swings one way, then its reverse. Amplitude is not volume. While amplitude is an objective measurement, volume is the subjective perception of how intense or loud a sound is. While a typically higher amplitude is perceived as higher volume, there is no consistent measured correlation between them across humans. Acoustically, volume is frequency-dependent. Frequencies at the extreme upper or lower end of the spectrum of human hearing require more amplitude, logarithmically, to achieve the same loudness than that of a frequency in the center of the typical human hearing range.

Onstage, the cellist is still tuning to the piano. If you could observe a single molecule in that disturbance, what you would see is the air molecule propelled forward, colliding into the next molecule, and bouncing back in the opposite direction, only to collide into the molecule behind it. It bounces forward and back again and again. The air molecule oscillates in this way until it comes to rest. The air molecules that are disturbed by the resonating body of the cello collide with more force than those disturbed by the string alone. That force is amplitude.

Now the pianist gives the cellist a G, and the cellist moves down to the G string, creating another sphere of disturbance, and that disturbance travels all over the theatre. You sit in your seat, you listen to her tuning, and you can tell the difference between the G note from the piano and the G note from the cello.

The vibration of air that the G string created hits your outer ear, called the pinna, which directs the air vibration into the ear canal, where it affects the eardrum. The eardrum (tympanic membrane) is elastic. When the air vibration applies positive pressure (compression), the eardrum stretches forward, deeper into the middle ear. The middle ear is filled with air. The eustachian tube is a tube that connects the middle ear to your throat, equalizing the air pressure. Three tiny bones, the malleus, incus, and stapes (together called the ossicles), live in the middle ear, and are suspended by ligaments so they can vibrate. When the eardrum oscillates, it causes the bones to vibrate. The stapes has a flat plate structure called the footplate that is connected to the oval window of the cochlea. Amplification of the vibrations occurs as the result of the tympanic membrane's larger surface area and the lever action of the middle ear. That amplification can be measured in the very small oval window. The cochlea is shaped like a snail shell and is filled with fluid. Inside the cochlea is the basilar membrane, anchored to the inner walls of the cochlea at two points along its edges and is stretched across the cochlea's interior like a ribbon microphone. The cochlea is filled with fluid. The basilar membrane vibrates in response to the cochlear fluid, which is triggered by the vibration of the oval window. The basilar membrane has a varying width and stiffness in places, which allows it to respond to both higher frequencies and lower frequencies. The basilar membrane supports the organ of Corti, which contains hair cells. When the membrane vibrates, it makes the hair cells vibrate. This is where the kinetic vibrations are transduced into electrical signals utilizing the stereocilia, which sit at the top of each hair cell. The electrical signals are carried to the brain via the respective auditory nerve (one nerve for each ear) to be processed. This is an analog electrical signal that matches the frequencies of that G string and the harmonics it creates.

Harmonics are additional sine waves that the original G note, called the fundamental, creates. Harmonics are related to the fundamental frequency (or fundamental pitch) at simple frequency integers. The fundamental is usually the wave with the strongest amplitude, and when combined with harmonics forms the complex sine wave that reaches your ear. All musical

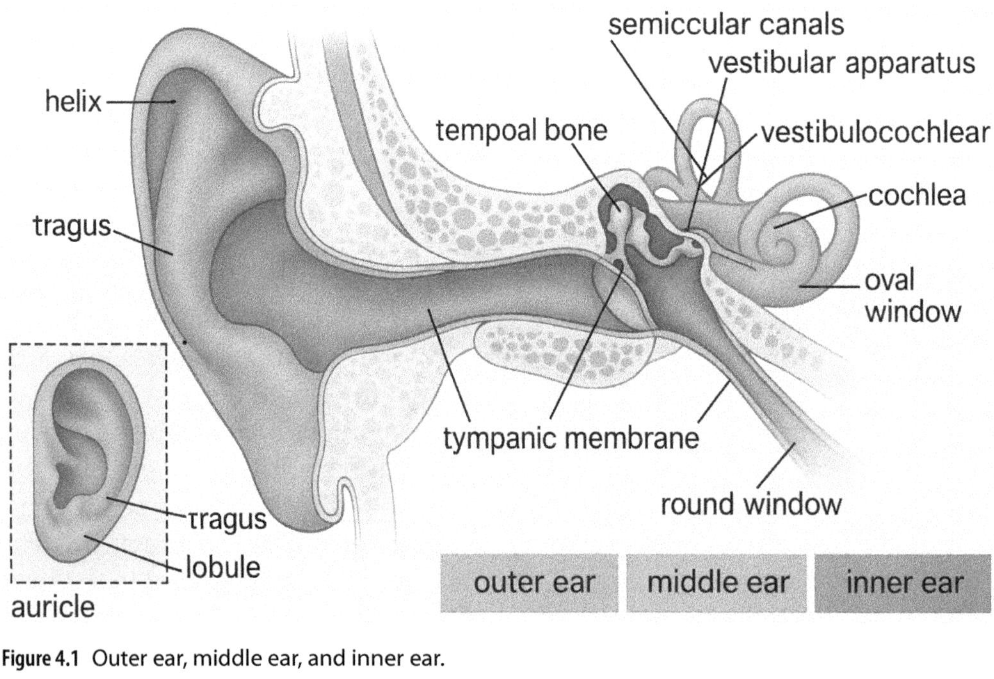

Figure 4.1 Outer ear, middle ear, and inner ear.

Drawing by the Lana Floriana.

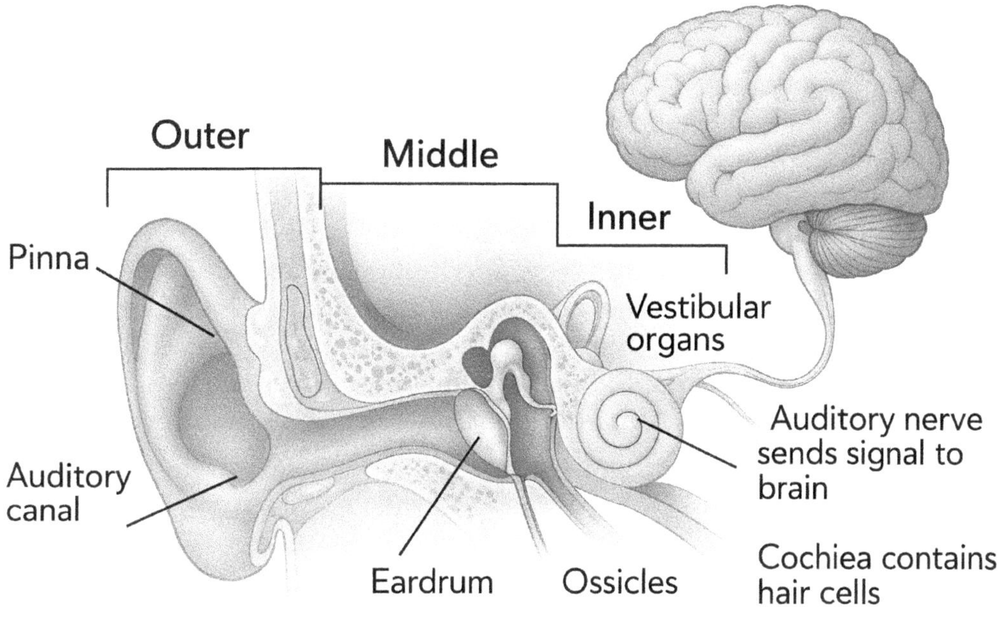

Figure 4.2 Auditory nerve.

Drawing by the Lana Floriana.

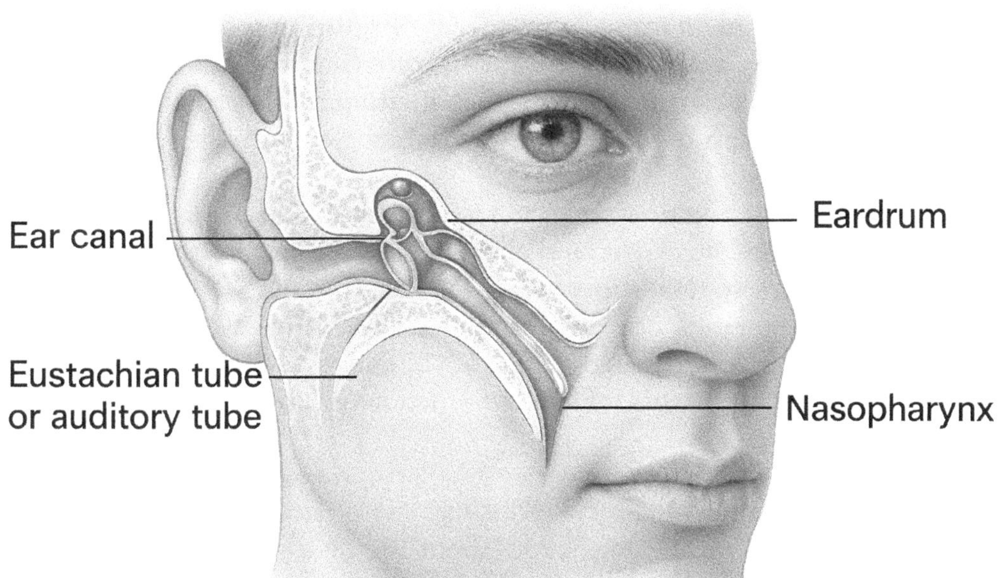

Figure 4.3 Eustachian tube.

Drawing by the Lana Floriana.

notes are complex sine waves, that is, a combination of the fundamental and its harmonics.

The unique timbre (TAM-ber) of every sound is a result of the specific mix of the sound's harmonic frequencies combined with how the amplitude of each harmonic relates to the others and shifts over time. When the pianist plays the G, it has a different quality of sound than that of a G of the same octave played on the cello. We can easily differentiate between the two. A trombone's timbre and, say, a xylophone's timbre are completely different, even though they may play the same note. Timbre is how we perceive the specific and unique mix of the amplitude of each harmonic frequency that the pitch of the instrument creates over time. Every sound has timbre: the knocking on a door, the note on a flute, and even human vocalization such as speech.

How Humans Perceive Sound

Humans have two auditory nerves, one from the left ear and one from the right. The auditory nerves stimulate, or innervate, the brain stem, which sends information in the form of electrical signals to several different parts of the brain, and particularly the auditory cortex. This enables us to perceive all kinds of information about the sound we hear. Humans can localize precisely where the sound is coming from. We can also perceive the duration of sound. We can process sounds with dynamic sound patterns that change over time in terms of frequency, amplitude, rhythm, etc. A cardinal's song, for example, consists of trills, pitch changes, random tempos, changes in loudness, and melodies that fluctuate. We perceive all those changes. Humans can determine timbre, which we know is a function of complex sound waves.

Auditory processing does not seem to happen in specific locations of the brain *per se* but does employ neural circuits. Neural circuits are a network of neurons that work together to process information and perform a specific function. The larger neurons can be connected to multiple neural circuits, contributing to different functions for each circuit. Scientists have identified many specific neural circuits for specific functions relating to interpreting the sound we hear.

Neural Circuits Involved in Auditory Processing

The immense amount of information about the neural circuits involved in processing the electrical signals that represent sound are outside the scope of this book. Furthermore, there is still much we do not know about how the brain interprets these electrical signals. The following is an attempt to give you, the reader, a simple roadmap of the basic path those signals take through the brain's auditory regions. You can think of this as simply signal flow, with simple descriptions of the functions of each neural circuit. If you are interested in this information in more detail, I encourage you to take the time to dive into the plethora of medical and scientific literature available.

In brief, the auditory cortex processes information in stages, and each stage consists of its own neural circuits:

First Stage: The primary auditory cortex, rostral area, and rostro-temporal area are named the auditory core areas. They are part of the first stage of processing. They perform tonotopic processing that involves pure tones with precise frequency encoding. They are the primary entry point for auditory information, extracting frequency-based features and distributing them to the belt for further processing. Their parallel thalamic inputs and interhemispheric connections ensure robust initial processing.

Second Stage: The parabelt region is next. It involves higher-order processing, such as non-tonotopic processing, such as auditory patterns, sound localization, and object recognition. It distributes it to the multimodal (involving multiple types of information) and cognitive regions of the brain.

Third Stage: Additional neural circuits are in the brain's temporal regions, parietal regions, and frontal regions. It is in these higher areas of the brain where sound information, visual information, and somatosensory information converge for tasks such as more precise sound localization and audiovisual integration. This includes auditory functions, such as speech processing and other complex dynamic sound patterns.

The implications of this knowledge are that the brain employs both serial processing and parallel processing. Serial processing is a stricter hierarchy. Sound information travels a distinct path from the core to the belt, then to the parabelt, and then to higher regions of the brain. Parallel processing allows independent processing in separate regions, providing redundancy. Stream specializations imply specialized circuits for different auditory tasks. Finally, neuroplasticity (the brain's ability to adapt and reorganize itself) allows neurons to adapt their response characteristics, such as responding to familiar tones or the reconfiguration of their connections to adapt to changing auditory inputs after repeated exposure to sound frequencies. Neuroplasticity allows a musician to learn subtle pitch differences or adjust their firing thresholds in noisy environments, effectively "turning down the input gain." Neuroplasticity is what allows someone to be in a crowded, noisy room and hear their child crying in the next room. Neuroplasticity even allows us to compensate for hearing loss in one ear or both.[1]

Table 4.1 Auditory Processing Neural Circuits.

Region/Group	Stage/Level	Key Functions	Processing Types	Connections/Implications
Auditory core areas (primary auditory cortex, rostral area, rostro-temporal area)	First stage (primary entry)	- Tonotopic processing (pure tones, precise frequency encoding) – Extract frequency-based features	Serial (initial hierarchy) Parallel thalamic inputs	Signal flows from the auditory nerves, then to belt regions; interhemispheric connections for robust processing; enables redundancy
Parabelt region	Higher-order	- Non-tonotopic processing (auditory patterns, sound localization, object recognition)	Serial (follows core) Parallel streams for specialization	Distributes signal from auditory core areas to multimodal/cognitive regions; supports task-specific circuits
Higher neural circuits (temporal, parietal, frontal regions)	Advanced integration	- Convergence of auditory, visual, somatosensory info – Precise localization, audiovisual integration – Speech processing, complex dynamic sounds	Serial (end of hierarchy) Parallel (independent multimodal paths)	Distributes signal from parabelt region. Enables complex tasks; neuroplasticity for adaptation (e.g., pitch learning, noise filtering, hearing loss compensation)

Table by the author.

How the Brain Chooses Meaning From Sound

In *The Predictive Mind*, Jakob Hohwy poses the idea that our perception is a form of causal inference, where our brain infers the causes of sensory input. Our brain is contained inside our dark, quiet skull, and the only way our brain understands the outside world is through our senses. In our complex world, where multiple causes can produce similar effects, reliable inference requires some constraints. Our brains choose the most probable cause by combining current sensory input and prior beliefs.

> *Consider this very simple scenario. You are in a house with no windows and no books or internet. You hear a tapping sound and need to figure out what is causing it.*
>
> *This illustrates the basic perceptual task. You are like the brain, the house is the skull, and the sound is auditory sensory input. As you are wondering about the cause of the input, you begin to list the possible causes of the input. It could be a woodpecker pecking at the wall, a branch tapping at the wall in the wind, a burglar tampering with a lock, heavy roadworks further down the street, a neighbor's loud music, or those kids throwing stones; or it could be something internal such as loose water pipes banging against each other. Let your imagination rip: it could be that your house has been launched into space overnight and the sound is produced by a shower of meteorites. There is no end to the possible causes. Call each of these possibilities a hypothesis. The problem of perception is how the right hypothesis about the world is shaped and selected.*[2]

Hohwy argues that the mind is fundamentally a hypothesis-testing mechanism that perceives and interacts with the world by minimizing prediction errors. Again, these predictions are based on current sensory input and beliefs learned from prior experiences, and the brain chooses the most probable explanation. We react to these predictions with action as well as autonomic bodily processes, such as increased heart rate or the release of epinephrine or dopamine.

How Emotions Come to Be

Researchers these days reject the traditional view of emotions. Throughout history, scientists believed emotions to be universal and hardwired into us, a response to triggering events. In her groundbreaking book *How Emotions are Made*, Lisa Feldman Barrett proposes a different theory, based on her years of research in neuroscience. She proposes her theory of *Constructed Emotion*,[3] which argues that emotions are not triggered at all, but are constructed by the brain through predictive processing. The theory of predictive processing posits that the brain's predictions are based on current sensory input added to prior beliefs formed from experience. Barrett adds cultural concepts to that recipe, which implies that emotions vary across cultural boundaries.

Defining What We Call Emotion

There are countless emotions we are capable of feeling, especially if you take into account that the emotions we feel and how we express those feelings vary from culture to culture. Our brain catalogs emotions into diverse groups of instances that vary widely. At the top are two main qualities, which Barrett describes as valence and arousal. Valence is the degree to which the experience is positive or negative, referring to how pleasant or unpleasant it feels. Arousal describes the intensity of the emotion being experienced, ranging from low (calm, relaxed) to high (excited or tense). An emotion like fear might have an unpleasant valence if someone is threatening you. Then there is the kind of fear you may feel watching a scary movie, which can be fun. In that case, the valence might be more pleasant. The arousal or intensity would depend on how intensely you feel afraid.

Fear is a category. For example, there is the fear of heights, the fear of missing a deadline, the fear of being turned down for a date, and the fear of an unexpected noise in the house. On the other hand, there is the fear of riding a roller coaster, walking through a haunted house, skydiving, bungee jumping, and playing a scary video game. Anger is a category. Happiness is a category. Sadness is a category. Our brains are wired to learn these categories based on current sensory input and what it has learned from past experiences.

According to Barrett, emotions start with interception, which is the brain's awareness of the current bodily state. The brain is constantly sensing internal physiological states such as heart rate or body temperature. As a predictive organ, the brain uses past experiences to anticipate bodily needs. Emotions spawn from concepts – mental categories learned through experience and shaped by culture. These concepts are based on goals, organizing sensory input to achieve specific outcomes, like survival or social connections. She explains that emotions require perception shaped by social context, like money or laws. These emotion concepts prescribe actions and communicate meaning within a culture. Thus, she argues that emotions are not just personal, but social tools, influencing behavior and relationships through shared understanding.[3] In How Emotions Are Made, she writes:

> *An emotion is your brain's creation of what your bodily sensations mean, in relation to what is going on around you in the world. In every waking moment, your brain uses past experience, organized as concepts, to guide your actions and give your sensations meaning. When the concepts involved are emotion concepts, your brain constructs instances of emotion. If you're in a dark alley and your heart is racing, your brain might predict*

"Fear" and so you feel afraid. The same sensations in a different situation, like running in a park with your friends, might cause your brain to predict "Excitement," and so you feel excited.[3]

How Will You Apply This Knowledge to the Craft of Sound Design?

Now that you understand the brain's predictive nature in creating emotions, you can use that to create meaning. You must remember, however, that each person is unique and interprets in their own way, and now you know why.

It is a losing battle for a sound designer to set their goal on making an audience feel any specific emotion. This is manipulative. It is asking for the joke and not the tea. That is why it is better to present information about the story to include sound or music that represents what the character is feeling but not saying. To do that, the sound must be meaningful to you, and that is the best measure to base your choices on. If a sound you have chosen creates the same meaning in members of the audience that it has created in yourself, then that builds a sense of connection between them. A sense of community is created with each other, a community in which you are a part.

Predictive processing suggests the brain actively predicts sensory input, including sound, based on internal models, minimizing prediction errors to form perceptions. For sound designers, this means audiences are not passive recipients but active interpreters, whose prior experiences and expectations shape how they perceive your sound design. Knowing that listeners' brains anticipate sounds (e.g., expecting a crescendo in a dramatic score) allows you to manipulate these predictions to create surprise, tension, or catharsis, enhancing emotional impact. For example, in a noir play, you might want to create a feeling of tension and resolution. You might start with building a predictable pattern, like repeating sound effects or musical phrases – maybe the sound of hail on the window of the front door. Then introduce something unexpected such as someone outside pounding on the door. In the case of music, you can repeat the same phrase until finally, at just the right moment, the music resolves.

Sounds influence autonomic nervous system responses (e.g., heart rate, arousal), as the brain infers meaning from auditory cues. For example, a sudden loud sound may trigger a startle response, while a soft lullaby can be calming. Understanding these responses helps you design sounds that evoke specific emotions or moods, such as fear in horror films or tranquility in meditation apps, by aligning with the brain's predictive and emotional processing.

Context shapes meaning. The brain uses context to disambiguate sounds, where a sound's meaning depends on contextual cues (e.g., a tap could be a woodpecker or a burglar). Predictive processing emphasizes that prior beliefs, environmental context, and cultural learning all influence auditory perception. Sounds paired with visuals can amplify their perceived reality or emotional weight.

Sound effects can effectively support a feeling of plausibility – environmental sounds in a street scene, the sounds of a lazy river in the evening, or in the Amazon Jungle. You might use specific sound frequencies to trigger autonomic responses, such as low rumbles or high-pitched birdsong.

You might create ambiguous sounds that invite the listener's brain to predict meaning. You can create a metaphor for wind with a mix of tones – trumpets for voices and popping bubbles for machinery. The audience will perceive these sounds as what you intended, thanks to their brain's ability to resolve ambiguity based on contextual cues and past experience.

I urge you to study psychoacoustics to learn how frequency, timbre, and spatialization affect perception to design sounds that exploit the brain's auditory processing functions. Use dynamic range strategically, collaborate with visual artists, incorporate cultural context, and experiment with silence.

Understanding how humans receive, process, and derive meaning from sound through predictive processing empowers sound designers to create experiences that are perceptually engaging, emotionally resonant, and narratively coherent. By manipulating predictions, leveraging context, evoking autonomic responses, guiding attention, and integrating multisensory elements, you can craft soundscapes that deeply connect with listeners. Applying these principles, grounded in Hohwy's framework, enhances your ability to shape auditory experiences in films, games, installations, or music, making your work more impactful and memorable.

Chapter 4 Exercises

Exercise 1: Sound Wave Manipulation

Objective: Practice manipulating sound wave properties to influence emotional perception based on the brain's processing of frequency and amplitude.

Instructions:

1. Record a clean, single-tone sound in an audio editor.
2. Use the editor's pitch-shifting tool to create a version 10% higher and 10% lower in frequency.
3. Use the gain or amplitude adjustment to increase the energy of the original sound by 3 dB.
4. Journal about the feeling you experienced when the sound was changed.

Exercise 2: Contextual Sound Design

Objective: Experiment with creating soundscapes that shape narrative meaning through context, leveraging the brain's predictive processing.

Instructions:

1. Create or choose a 30-second soundscape for a short scene (e.g., a character walking through a forest).
2. Design two versions of the soundscape: One suggesting calm and one suggesting danger.
3. Select or create a neutral soundscape.
4. Ensure each soundscape uses similar frequency ranges but differs in timbre and rhythm to reflect safety or danger.
5. Journal your analysis on how context shapes meaning, referencing predictive processing from the chapter.

Exercise 3: Timbre Exploration

Objective: Explore how timbre influences emotional response by comparing sounds with distinct harmonic profiles.

Instructions:

1. Choose three instruments or sound sources with distinct timbres (e.g., a flute, a drum, a human voice).
2. Record the same note or sound from each source.
3. Arrange the sounds in an audio editor to create a one-minute composition that cycles through each sound, with a few seconds of silence between transitions.
4. Play the composition for three–five listeners and ask them to describe how each timbre makes them feel.
5. Write a reflection on how timbre, as a function of harmonics, affects emotional response.

Exercise 4: Predictive Processing Experiment

Objective: Apply predictive processing principles to design a sound sequence that manipulates audience expectations.

Instructions:

1. Design a one-minute sound sequence that builds a predictable pattern and then introduces an unexpected element.
2. In an audio editor, create a looping sound pattern (e.g., ten seconds of steady footsteps at 120 BPM).

3. After three loops, introduce an unexpected sound (e.g., a loud door slam or a high-pitched screech).
4. Play the sequence for five peers and ask them to describe their physical and emotional reactions.
5. Write an analysis linking listener reactions to predictive processing and autonomic responses, referencing the chapter's discussion of the brain's prediction errors.

Additional Reading

- *How Emotions Are Made: The Secret Life of the Brain, Lisa Feldman Barrett*, Houghton Mifflin Harcourt.
- *The Predictive Mind*, Jakob Hohwy, Oxford University Press.
- *Seven and a Half Lessons About the Brain*, Lisa Feldman Barrett, Houghton Mifflin Harcourt.

5

Sound and Noise

For our purposes, *sound* is everything the audience hears that the sound designer intends for them to hear, while *noise* is everything the audience hears that the sound designer did not intend for them to hear.

If a tree falls in the forest and nobody hears it, does it make a sound? Scientifically, of course it does. Sound is generated from the movement of the leaves, the cracking of the wood, the crunching of branches as they are crushed by the trunk, and the impact sound of the trunk hitting the earth. Those sound waves are within the spectrum of human hearing, so yes, it makes a sound. Sound is created. Sound happens.

In the theatre world, if a tree falls onstage and the audience hears it, does it make a sound? That would depend on whether the sound designer intends for the audience to hear it, and if so, then yes, it makes a sound. If the sound designer does not intend for the audience to hear the sound the tree makes when it slams onto the deck, then it is noise. Do everything, and I mean everything, possible to get rid of that noise.

Music from the pit is sound. The actor's high heels clicking across the stage could be a sound in a play I have designed. A door slam on cue could be a sound. An actor's dialogue is sound. An actor drumming their fingers on a table, shattering a glass against the wall, or laughing loudly are all examples of sound. If it is intended, it is sound.

A car with super subs at volume 11 passes by the theatre. That would be noise. An incandescent lamp in a lighting instrument exploding in the middle of a scene is noise. A stagehand accidentally drops a tray of silverware just off-stage.

DOI: 10.4324/9781003519393-7

The crash you hear is noise. The mixing engineer fails to turn off an actor's mic when they exit, and we hear the actor's voice in the house. That is noise. An actor burps in the middle of a song. . . . well, that is worth talking about.

There are times when an inadvertent vocalization of an actor is desirable and therefore could be sound if the sound designer wants it to be heard. On the other hand, the sound designer might intend not to hear anything coming out of an actor's mouth except their scripted lines and/or lyrics. For that matter, if an actor is playing drunk, an ad-libbed burp might be exactly aligned with the designer's intention. For all intents and purposes, a burp could very well be sound.

Types of Theatre Sound

There are several classifications, or types, of sound in the context of theatre. Understanding the differences will help you use them more effectively.

Real Sound

Real sound is actual sound created live, and not an effect. It includes:

- Actors speaking, singing, or otherwise vocalizing.
- Sound that occurs as a result of the director's blocking (action) such as footsteps, door slams, or throwing a baseball through a real plate glass window.
- Music played by an actor onstage with a real instrument.
- Music or accompaniment coming from an orchestra that is not reinforced. Even with the use of microphones, a portion of the audience will hear the orchestra acoustically, along with what is coming out of the speakers.
- Off-stage sound, if it is the real thing. For example, an actual tray of dishes thrown to the floor by a stagehand. Another example would be if the blocking requires an actor to be digging in the dirt onstage, the actor might simply indicate that action with a shovel while a stagehand thrusts another shovel into a box of real dirt off-stage.

Processed Sound

This is actual sound that you have manipulated in some way such as:

- Amplified sound reinforcement, including real sound that occurs off-stage that you reinforce.
- Sound that you process, but not necessarily amplified. For example, a pitch change, distortion, a reverb effect, or other real-time effects.

Sound Effects

A sound effect is a sound that represents a different sound. It has the effect of sounding like something else. You might say that it is artificial because it is created by different means. Sound effects are used in plays, movies, and other broadcast productions.

Performed Sound Effects

These are techniques used to create live sound effects. (I have often heard these referred to as *Foley* effects, but that would be inaccurate. Foley is a term for a technique developed by Jack Foley and is primarily associated with recorded sound for film post-production, not theatre.) Some examples of performed sound effects are:

- A slapstick.
- A drum that sounds like thunder.
- A bowling ball rolled in a track above the theatre, known as a thunder track.

Recorded Sound Effects and Music

Recorded sound effects are simply sounds that have been recorded and played back on cue. It can be real or a sound effect, as long as it is recorded. The same goes for pre-recorded music.

Silence

In the last chapter, we defined sound as everything that the audience can hear that the sound designer wants the audience to hear. However, there is one thing in the sound designer's toolbox that the audience cannot hear. Silence. The absence of sound. You cannot hear what is not there.

Silence is the absence of sound. It is the space between the end of one sound and the beginning of the next. Carefully chosen, silence can help create a profoundly meaningful moment.

Here on Earth, no one will ever be in an environment with absolute silence. Even if in an anechoic chamber, acoustically isolated from the rest of the world, people have reported hearing their own heartbeat and blood rushing through their veins.

For the purposes of the dramatic, it takes sound to experience the illusion of silence. When an audience thinks they are experiencing silence, they are experiencing the feeling of silence more than actual silence. A few examples:

- The quiet space between two loud sound cues will feel silent in contrast.

- Two actors are onstage talking and experience a moment of awkward silence. You can create this feeling by increasing the volume of a bird tweet and placing it smack dab in the middle of that awkward silence.
- Things feel "quiet" when it snows because the snow absorbs sound energy. It makes the environment dry (less sound reflection). You can mimic that effect in a dramatic moment simply by piping barely audible pink or brown noise through the PA system.

Other Qualities of Sound

Sound and Time

Have you ever watched a movie where a character snaps their fingers and stops time for everyone around them except for the two main characters? Everyone else stops, frozen at the task they were performing. Perhaps a bullet stops, suspended in mid-flight. The two characters talk, plan, and escape the danger. The character snaps their fingers, and time begins again.

Of course, if that were to happen in the real world, there would be no sound. Sound has speed, which is a factor of distance over time. There would be complete silence, as the molecules in the air (or other medium) would be still. For that matter, there would be no light.

To exist, sound requires time. Without time, there would be no sound. Every sound in a production has a beginning, a middle, and an end. Another way to explain that would be that there is a moment before the sound begins, the moment the sound begins to propagate, a series of moments that sound is active and changes, and a moment after that sound ceases to exist. Remember, sound is not energy. It is the transfer of energy.

Once it begins, all sound changes over the course of its life. It ebbs and flows, based on a million circumstances that affect it, like air currents, air density, and humidity in the air as the signal flows toward your ear. Those changes happen during the construct of time.

Thinking about sound as existing in time and changes over the life of the sound will help you understand that silence does not exist in time. Silence exists outside of time. You might think that the life of silence exists only in the space between the end of one sound and the beginning of the next, but that is not accurate. Silence always exists. It exists when there is no sound, and it exists when sound is present. Underneath the sound. Even as the air molecules bang into each other in all directions, passing energy down the line, there is space between the air molecules. In that space, there is silence.

This way of thinking has utility for the sound designer. We will explore that further as we talk about the sound designer's tools in another chapter.

Sound Does Not Have Intrinsic Emotional Qualities

Some believe that sound has intrinsic human emotional qualities, but that is not true. The sounds we hear do not inherently carry human or emotional characteristics. This is a misconception that encourages you to control the outcome by manipulating the audience. Trying to elicit an emotional response by playing a sound that reminds you of, say, an infant or a playful puppy is a frustratingly losing proposition.

Sound can, however, trigger a memory in an audience member, reminding them of something else, like a baby, a mother, or a puppy. The sound itself carries no inherent meaning, resonating only with audience members who associate it with a personal memory or experience.

Even if the audience makes the engram connection you hope they will make, there is no guarantee that the audience member will experience the same emotional reaction you were hoping would occur. The meaning an audience member derives from a sound is wholly shaped by the narrative or memory that the sound evokes in them.

When you add sound, do not expect that the audience will respond in the same way you do. That is an unachievable goal. It is an unrealistic expectation. Add sound to inform the audience by providing some information and allow them to experience what they will.

Sound Is Movement

A signal is a representation of sound that travels. A few forms it can take are:

- An electrical signal through a wire.
- A kinetic impulse, such as the oscillating molecules like air molecules, or some other medium such as water.
- A magnetic impulse such as a tape recording.
- The computer-coded format of a computer audio file that can be translated into sound.
- An electromagnetic wave, such as a radio wave.

Another way to look at a signal is to transfer information by motion, action, movement, or sound. We are going to combine all of this into the definition for the sake of a common language.

Signal flow is the movement of the signal down the path that sound takes from its creation to landing on human ears. Along that path, the signal takes many forms, as mentioned earlier. A singer's vocal cords create the sound that travels through the air, through the transducer, through the cable, through the mixing console's preamps and analog/digital converters (if it has not been converted to digital before that point, in which case it would travel through a computer network), through the mixer channels, out to the cables, through the amplifiers, through the speaker cable, out the speaker transducers, through the air again, and all the way to the ears (Human, or otherwise). The listener's brain interprets the signal and finds the performance meaningful.

Always return to the signal flow. Understanding its path in each venue and production is going to help you process and manipulate it with intention.

- It is going to significantly speed up troubleshooting, in the event something goes wrong, such as distortion, drop out, system noise, or feedback.
- It is going to help you make decisions on how to manipulate the signal in a dramatic or interesting way. For example, let us say you have decided to distort your signal during a specific moment in the play. You can run a signal through a distortion effect processor, or you may decide to overdrive the compressor creating a distinct type of distortion. You may decide to distort it in some other way, depending on the play.
- It is going to help you quickly shape live sound by selecting your gain structure and deciding in what order to place the signal processors such as noise gate, EQ, and compressors. The same EQ curve will sound different depending on whether it follows the compressor or precedes it.

Signal flow is the foundation of all your artistic decisions that have to do with your delivery system. It is the foundation of all your mixing decisions, all your troubleshooting decisions, and all the solutions you produce because of that troubleshooting.

Noise

Noise is everything the audience can hear that you do not want them to hear. Noise can distract an audience from fully focusing.

- It takes the audience out of the world of the play. It reminds them that they are watching a play.
- It undermines the focus you would like to put onto a sound cue and the moment.
- It can sometimes actually stop the play or force a pause. I once designed an outdoor production of *Twelfth Night*, and the stage was

adjacent to a hospital's emergency helicopter pad. When that helicopter landed, there was nothing for the actors to do but stop the play and wait until the helicopter took off again.
- Noise can drown out an actor's dialogue by making it challenging to obtain vocal intelligibility, which makes it impossible to follow the story.

You must do everything you can to eliminate noise to deliver your sound in an optimal and pleasing way.

Noise comes in various ways that stack. Any one of the following examples has a volume of energy, and more than one combined simply increases the loudness of the overall noise.

- Noise of the building itself, such as HVAC, noisy radiators, or pipes.
- Outside noise, such as a thunderstorm directly over the theatre, traffic, sirens from emergency vehicles, anything that is incidental, and outside the theatre or audience area (in the case of outdoor theatre).
- Audience noise such as talking, unwrapping candy, or a baby crying. Applause is sound since you will want the audience to hear itself applaud.
- Sound system noise, such as amp noise, or noise coming from the monitors.
- Noise from backstage, such as the movement of scenery or whispering. Hopefully, the stagehands know to be quiet and never talk loudly enough for the audience to hear them.
- Stage noise is noise that comes from the stage you do not want to hear, like an onstage band that is picked up by actor microphones or a creaky set piece.

If your noise floor is high, it often comes from a few different sources and not just one. The only way to deal with that is to address one source at a time the best you can. Perhaps first you will address the HVAC, and then eliminate as much noise from the sound system and then address backstage noise. The more layers you strip away, the better you will be able to identify other sources.

There is a common assumption that you can mask noise. For example, if from the audience area you can hear the onstage monitor wedges facing musicians, a common impulse is to make the sound system even louder. Then we wonder why people complain that the show is too loud. You cannot "unhear" existing noise, even when you add other sound (or noise) to it. Noise stacks. In the same way, it is unwise to expect an audience to ignore the noise. It all combines into the overall "sound" of the show. Eliminate the monitor noise if you can. If you cannot, suppress it as much as possible to limit its ability to compromise the sound you want them to hear.

Deliver sound. Eliminate noise.

Chapter 5 Exercises

Exercise 1: Sound and Noise Observation in a Live Performance

Objective: Identify and differentiate between intentional sound and unintended noise in a live theatre setting.

Instructions:

1. Attend a live theatre performance or rehearsal.
2. During the performance, note all auditory elements in two categories: Sound (e.g., actor dialogue, on-cue door slams, orchestral music) and noise (e.g., audience coughing, backstage clatter, external traffic).
3. Create a table listing at least five examples of each, including a brief description of the context (e.g., "Actor's mic picking up costume rustling during a quiet scene").
4. Write a 150-word reflection on how one specific noise disrupted the audience's immersion and propose a practical solution to eliminate or mitigate it (e.g., soundproofing, mic gating).

Exercise 2: Crafting a Moment of Silence

Objective: Design and implement a moment of perceived silence to enhance a dramatic scene.

Instructions:

1. Select a scene from a play with a moment of tension, pause, or transition.
2. Create a sound cue that simulates silence using subtle sounds (e.g., faint pink noise to mimic a snowy environment or a single bird tweet during an awkward pause).
3. Rehearse the scene with a small group, integrating your silence cue using basic theatre sound equipment (e.g., a PA system or portable speaker).
4. Collect feedback from peers on how the silence affected the scene's mood or impact.

Exercise 3: Classifying Sound Types in a Theatrical Scene

Objective: Categorize sounds in a live theatre scene according to the chapter's sound types.

Instructions:

1. Choose a scene from a play with diverse auditory elements.
2. Create a chart classifying each sound as real, processed, or effect-based.
3. For each sound, note its purpose in the scene (e.g., "Real door slam reinforces action realism").
4. Present your chart to the class, explaining one example from each category and how it supports the production.

Exercise 4: Signal Flow Mapping and Troubleshooting

Objective: Diagram and analyze signal flow in a theatre sound system to troubleshoot potential noise issues.

Instructions:

1. Visit a theatre's sound booth or use a provided diagram of a typical live theatre sound system (e.g., microphone, mixer, amplifier, speakers).
2. Draw a rough but detailed signal flow diagram, labeling each component from sound source (e.g., actor's voice via mic) to output (audience's ears).
3. Identify one potential point in the signal flow where noise could be introduced (e.g., a loose cable causing static, an unmuted mic picking up backstage chatter).
4. Describe a troubleshooting step to address this noise issue, referencing the signal flow.

Exercise 5: Exploring Sound's Neutrality Through Audience Perception

Objective: Investigate how sounds trigger audience memories or associations without inherent emotional qualities.

Instructions:

1. Select three neutral sounds relevant to theatre (e.g., a creaking floorboard, a ticking clock, a distant bell).
2. Play each sound during a class or rehearsal setting for 5–10 peers, asking them to write a brief description (one to two sentences) of the memory, image, or story the sound evokes (not an emotion).

3. Compile responses in a table and analyze patterns in how audiences interpret the sounds.
4. Write a 200-word reflection on how these findings align with the chapter's assertion that sound lacks emotional qualities and relies on audience perception, discussing implications for sound design choices.

6

The Sound Designer's Toolbox

As a sound designer, you can use any tool available to you that will create and deliver the sound you want. A tool is anything that exists or will exist that helps you create sound. Go crazy. Use wet washcloths or the inside of a trash can or an airplane wheel or your neighbor's pet goldfish.

Some of the following tools are practical, like a speaker array, a crash box, or a mixing console, and others are conceptual, such as understanding the difference between sound and noise. This chapter is not exhaustive. It is not meant to be. Your tools are limited only by your awareness of what is out there, as well as your own imagination.

The Sound Designer's Technique

These days, the sound designer's medium is a technically oriented one, but it was not always that way. All art forms have utilized the new technology of the day. Paint and brushes were at one time considered cutting edge, and so was the modern piano, the slide rule and compass, and the pottery wheel. Yet the painter spends a lifetime mastering the use of the brush. The pianist does so with their instrument, as does the architect or the sculptor.

If technology is the nexus of scientific knowledge for practical purposes, the sound designer can use the latest technology in their design to tell the story they are trying to tell, but they do not have to. They can utilize

DOI: 10.4324/9781003519393-8

the methods of old if they feel that is more meaningful. There are literally thousands of years of historical techniques to choose from.

Tool 1: Sound Cues

In theatre, a cue is a prompt that triggers a specific action, line, or event during the performance. It can be a line delivered, a physical movement, a sound, a lighting change, or other prearranged signal that prompts the actors, crew, or technicians to perform their next task.

The cue for an execution of a sound is called by the stage manager (or other person acting in their stead), who is prompted by their own cue (an action, line of dialogue, or other event).

A sound cue is technically the sound itself or an entire sequence of sounds. The sound cue triggers the next action. The sound cue may prompt the actors, stage manager, or anyone else tasked with doing something cued by the sound. It is a cue of sound: A sound cue.

A sound cue is one in a list of sound cues or sequences, even if it is a list of one. They have a sequence in which they are executed. The stage manager, prompted by a precise and deliberate moment in the play, calls the sound cue. It is executed in concert with the sound and silence that come before and after it. It waits in a list of sound cues to play its part in the symphony of the whole. It is a tree in the forest. There is a thread of commonality between all of your cues: A progression or an arc that parallels the arc of the story.

As discussed, the purpose of a sound cue is to inform the audience. Geographic location, attitude, mood, and character personality are examples of data that can be shared by a simple underscore cue.

Sound cues can be sound effects, recorded effects, music, live or recorded voice-overs.

Live Sound Cues

Live sound cues are either real sound on cue or performed sound effects, again, on cue. They include live music cues, live voice-over cues, sound created live onstage by actors, or live sound effects such as a crash box or a thunder sheet.

A crash box is a sound effects device used to create realistic crash, collision, or breakage noises during live performances. It is typically a large wooden or metal box filled with breakable items like glassware, dishware, wood blocks, or other fragile objects. To use it, the crash box is either dropped from a height backstage for a dramatic impact sound or shaken vigorously for a subtler effect.

A thunder sheet is a thin metal sheet, often suspended to a pipe or frame by ropes, that creates the sound of rolling thunder when shaken or wobbled by an operator. Often, two or more thunder sheets are used together, depending on how loud or long the cue should be.

Playback of Recorded Sound Cues

This is any media that is recorded and played back on cue. These cues are sound that is recorded onto a media platform such as a record, a compact disk, tapes, sampled media, or computer audio files. They are played back on cue. You might do the recording and editing yourself, or you might download sound effects and/or music and edit them for your purposes. You might even use it without editing, if it already suits your purposes.

Characteristics of Sound Cues

Every sound cue has specific characteristics. Each has its own purpose. Put these into your own toolbox.

Diegetic and Non-Diegetic Sound

A diegetic sound cue is a term that originated in film. It is the sound that both the characters in the play and the audience can hear. It would include sound effects or music played in the scene. This sound exists in the world of the play.

A non-diegetic sound cue refers to a sound that only the audience can hear. Transition music, underscore, etc., played for effect but not part of the scene itself. It is not present in the action. Another term would be commentary sound. They would be voice-overs, sound effects added for dramatic effect, mood music, or underscore.

Volume

Volume, or loudness, is crucial. The wrong volume can immediately remind the audience that they are watching a program, which will take them out of the world of the play and back into their seat in the theatre.

You do not need volume to achieve clarity. Distributed sound systems work. The manipulation of the loudness of a sound cue at the right moment is an art of its own.

Context

We have talked about context in an earlier chapter. Let us expand on that a bit. A sound cue is an event that can only be fully understood in terms of the circumstances that form its setting. The setting is:

- The physical circumstances of place and time.
- Conditions of the environment, such as a snowstorm, or on a boat with calm or rough water.

- Emotional momentum such as after or before a series of humorous moments, or tense moments.

Context is crucial to the meaning that it can inspire. Let us look at a few examples:

A woman is in her house. Suddenly, there is a hard rapping on the window. Is she in the living room? Is she in the kitchen? The ambiguity of the location of a sound cue might have many different meanings without more contextual information.

What if the woman were older, say in her 50s? What if she were alone in the house and in the library? Does the library have one small window, or does it have a wall of large, flimsy French doors? Do those circumstances inspire a moment of apprehension or even fear? Even for just the moment? What if there is a blizzard outside and the electricity and phone are out? What if it is morning? What if it is late at night? Perhaps it is late in the evening, and there is a large fire in a large hearth? More information: What if we know she has a pistol in the pocket of her sweater? That might inspire a different tension, as if you are ready to use it if needed. Here, the context pours meaning into the sound cue of rapping on the window. But what if the context changed?

Different context: Same woman alone in the house and the same room – the library with a wall of large, flimsy French doors with foggy windows. Blizzard conditions and snow drifts are growing higher and higher. The electricity is out. The phone is out. It was 1950 in Montana, on a ranch. No one for miles. There is a pistol in the woman's pocket. Add one bit of information: her grandson, 10 years old, went out to gather some wood for the fire 30 minutes ago and has not returned. In this context, a rapping on the window might bring a moment of hope, even for just a moment, that the child has finally returned.

Let us change it up again. Same woman, same room, and same snowstorm. No lights or phone. Same fire in the hearth. Same ranch in Montana. This time, the woman is not alone. She is with her grandson, who is sweating with fever, asleep on the couch near the fire. There is no sign of the storm letting up. Suddenly, someone raps on the window. In this context, we do not have prior clues about who might be rapping on the window. The meaning in the cue is a mystery. With the sick child in mind, the rapping could be that someone has arrived to help. With the snowstorm, the rapping could be someone in need of help of their own. It could be a threat. It could be a blessing. It could be just an added complication to the situation. This ambiguity adds tension to the moment. What comes next will define the meaning of the rapping, but in the gap before that next moment, the context of the cue brings the tension of ambiguity. The brain struggles for the right prediction.

How would the meaning change if you saw red and blue flashing lights against the window just before you heard the rapping?

Keep in mind that the meaning we are representing is the meaning the sound cue has for the characters in the play and not the meaning we want the audience to derive from it. Let the audience observe the characters and understand what the characters are feeling to derive their own meaning.

Every sound cue has a context, whether that context is well thought out or not. The sound of a police radio in the distance can inspire fear in the character, or relief. The sound of a baby crying can inspire sympathy in the character or worry. The sound of a toilet flush can inspire laughter in the character or disgust. The meaning that a sound cue inspires depends on the context in which the sound cue event takes place.

Image of Origin

An image of origin is the illusion of the specific location from which the sound is coming. This is where you want the audience to believe or perceive the sound originates. The placement of where the actual sound comes from might be the same as its image placement, or it might come from elsewhere, often somewhere completely different. You are at liberty to think creatively. Use the upstage drop (cyclorama) or wall to diffuse sound by focusing a hanging speaker at it.

One good way to image a sound is to simply place a speaker where you want the sound to come from. If there is an old phonograph playing a song, place a speaker inside its box, or better yet, tie into the speaker that was placed into the machine when it was manufactured. If there is the sound of two men fighting in another room, place a speaker just beyond the door to the room, outside of the audience's sightline. If you want music coming from the apartment next door, place a speaker just on the other side of the flat, again, outside of the audience's view.

Another good way is to use two speakers in tandem to image the sound location. One way is to pan a stereo signal in various degrees left or right.

Panning takes advantage of our human binaural functionality and how the brain interprets that aural information. Close your eyes, stretch your right arm out to the side, and snap your fingers. Or better yet, have a friend snap their fingers to your right side. You can tell where the snap is coming from. Your brain interprets the information it receives from the information it gets from both ears. It perceives the fraction of a millisecond difference that the sound arrives at each ear, also known as the time differential. It can also perceive the volume differential in each ear. Your brain can perceive the difference in the dynamic range in each ear. If the snap is on your right side, then it will receive a wider dynamic range than the left ear. The same is true with other differential cues such as room reflections and even tone. The brain perceives and interprets all this information and instantly determines where the sound is coming from. You can speculate about the benefits of this in human evolutionary terms, and you can use it to your advantage to image sound.

Remember that imaging the location of the sound does not have to be real; it can be an illusion. Panning images a sound through volume differential. Imagine you are wearing a set of headphones and listening to a recording of a solo trumpet playing a good tune. If the trumpet is panned to the center, what is really happening is that the trumpet is playing through the left and right speakers of your headphones at an equal volume. If you pan a bit to the right, then the trumpet is louder in the right speaker than the left. What is really happening is not that the right side is amplified, but that the left side is lowered in volume slightly. The more you pan to the right, the more you lower the volume of the trumpet in the left speaker, until finally, when fully panned right, the trumpet is out of the left speaker altogether.

Panning can be utilized onstage with two speakers, left and right, but it is not as accurate as headphones. To take full advantage of imaging through panning, you must be sitting in the house precisely in between those two speakers, and you have to be inside the splay pattern of each. The more off-center or off-splay you are, the less effective panning becomes.

Another more effective way to image the source of the sound is to take that volume differential and add a time differential between the two. Imagine you are sitting in the theatre and there are two speakers onstage on the same plane, that is, the same distance from the edge of the stage. They are playing that happy trumpet. If you put a 0.5 or even a 1 millisecond delay on one speaker, say stage right, it will mimic the time differential, and the brain will be fooled into thinking the sound is coming from stage left. Now, that delay time between the speakers will have to be adjusted depending on how far apart the speakers are. If you move one speaker upstage a bit or downstage a bit, that delay time will have to be adjusted as well. For now, though, let us keep them on the same plane, playing the trumpet with a time differential. If you lower the volume of the stage left speaker – the speaker without the delay, it does not affect the imaged location. You can make the delayed stage right speaker louder and keep that image. You can raise the volume of the stage right speaker without ruining the imaged location up to a certain threshold, but if you raise the volume too much, you can simply raise the volume of the stage left speaker to minimize the volume differential and restore that image.

I often use this technique with actor microphones. Any experienced actor knows how to project to the back of the house with their voice, but often, especially in music, you will want to reinforce their voices. This technique takes advantage of the rule that you can "not hear" a sound if it exists at a volume above the threshold of silence. You can only direct focus. As an audience member, you can hear the actors' voices acoustically, along with the sound of their voices coming out of the speakers. Since the speakers are louder, it can be distracting to see the actor speaking on stage, but their voice is way up in the speakers. You can image their voices back into their mouths with a time differential. I will delay the microphones coming into the mixer, which allows the acoustic voice to arrive at the audience earlier than the speakers do, and voila! Actors and their voices are imaged in the same location. It works even though the speakers are much louder than the actors. You can calculate the delay time depending on how far the stage is from the audience and then tweak it by listening.

You can also use this technique with sound effects. Remember the speaker placed near the phonograph? What if the speaker must be so small that you cannot pump enough volume out of it to be heard at the back of the room? You can play the music out of the house speakers in tandem with the phonograph speakers and create a delay in the main speakers just for that channel. This images the source location of the phonograph quite effectively. Sound comes out of the phonograph for image purposes, and sound comes out of the house speakers for clarity and volume purposes.

Perspective

Perspective is the representation of what a sound effect sounds like from a chosen character's point of view. A punch sounds different to you depending on whether you are the one getting punched or the one doing the punching. A sound cue has a chosen perspective or point of view, or it has an accidental one. Perspective allows the audience to experience more fully what the character is experiencing. By delivering the sound that the character of focus is experiencing, Gestalt psychology tells us that the audience will fill in other sensory stimuli with their imagination, which can sustain their captivation with the moment. (Remember that the audience's sustained captivation in a play is built from moment to moment.) If a person is conked on the head, a sound effect helps the audience understand the character's pain, if they buy into it. An exaggerated door slam can help the audience understand a prisoner's loneliness.

There are more perspectives to choose from than there are characters in the play. Perspectives can be from:

- A specific character onstage who maintains primary focus, or status, of the moment.
- A character who maintains secondary status or focus in that moment.

- A meta-character – an observer which the other characters are unaware of but is observed by the audience.
- The audience itself.

The sound designer must also decide the proximity of the sound's origin to each chosen perspective, that is, in a character's peri-personal space or extra-personal space.

One's peri-personal space starts from inside their body and extends outside their skin but within their immediate reach. Hold out your arm, and if you can touch it, it exists in your peri-personal space. A person's extra-personal space begins just outside their reach and extends outward to the limits of their awareness. If you stand on a mountain and take in the vista, everything you can see but cannot reach is within your extra-personal space. If an ant crawls up your leg, that ant is in your peri-personal space. If you close your eyes and can feel your heartbeat, that heartbeat is in your peri-personal space. If your friend is talking to you from six feet away, his voice originates from an area inside your extra-personal space.

If the moment requires a sound cue of someone cracking their back, for example, that sound cue will have different aural qualities depending on the perspective you choose. You might choose the perspective of the person cracking their own back. That would sound different, and probably louder, then if you choose the perspective of a secondary character who is, say, 6 feet away. If you chose the perspective of the audience, that sound cue would sound different still.

A good example of allowing the audience to hear a sound cue from a particular character's perspective is in Tennessee Williams' *A Streetcar Named Desire*, where we are allowed to hear the music and gunshot scenario that continually plays over and over in Blanche's mind. None of the other characters can hear it, but Blanche does, and so do we.

Other good examples are voice-overs that reveal what a specific character is thinking. Or if a character perceives the voice of God in his mind.

Non-diegetic also has a perspective. That perspective comes from the audience observer, and you can place that audience observer in the seat they occupy, or you can represent the sound as if the audience were in the scene. Let us take an example. A kitchen area in a brownstone apartment. The woman who lives in the brownstone is cooking on the stove and hears screaming outside the window. In a rush, she drops her dish towel and does not pay much attention to where it lands – right on the burner. As she runs out of the apartment to help whoever is screaming, she forgets to turn off the burner. Now the apartment is empty. The dish towel bursts into flames with a "whoosh!" That "Whoosh" sound is the object of our discussion here.

No character hears that sound; therefore, it can be argued to be non-diegetic, so we might choose to represent the sound from the audience's perspective. However, an argument could be made that since it occurs within the world of the play, it could just as equally be diegetic, and we might choose to represent the sound from a character's perspective, or even silent as no one is in the room. As a sound designer, you have a choice to make for this scenario. Are you going to choose the audience's perspective – in their seats? Or are you going to simulate moving the audience into that apartment? That conflagration will sound a bit different in terms of volume and dynamic range depending on your choice. The information will be conveyed from either perspective, but which perspective will be more meaningful in terms of other chosen qualities such as context and style?

Another example of non-diegetic sound is a voice-over from an unseen narrator. Again, that sound would be represented from the perspective of the audience. Perhaps in this case, you choose to keep the audience in their seats and not simulate a different location. In that case, you can still choose to image that voice-over from the same location as the other characters' dialogue by delivering it through onstage speakers, or you can simulate it coming from everywhere by placing it, say, in the main speakers above the stage.

Classifications of Sound Cues

- House sound: Pre-show, entr'acte, curtain call, and walkout sound.
- Transitional sound: They transition the audience into a scene or out of a scene. They can also carry momentum from one scene to the next, out of one scene, and directly into another.
- Dramatic sound or scenic sound: Sound that is part the story itself, either directly or indirectly, diegetic or non-diegetic. They occur within scenes or can be extended in between scenes, thereby also serving as transitional sound.
- Atmosphere or soundscape: Background sounds. They can provide information about the location or time period. This can also be used in a dramatic way to communicate a character's mood or emotions.
- Direct sound: Sounds that are the result of a performer's action. The actor plays the guitar or drops a tray of dishes.
- Indirect sound: Sounds that are not caused by a performer's action. Car crash in the distance, a clock tower, or an airplane flyover.

- Voice-overs
- Designed sound effects: Sounds that do not occur in nature or are impossible to record in nature. The sound of a spacecraft or laser gun that has not been invented yet.
- Underscore: Underscore can be music or sound effects that the actors are not aware of. It is a musical interpretation of what the actor is feeling. It can propel the story, all by itself, by changing the context of a character's dialogue or action. For example, let us say that the character picks up and reads something that is not meant for their eyes. Perhaps a medical diagnosis meant for another character. The music can set a sinister context, telling the audience that the character intends to do something evil with that information, while different music might set a sympathetic context to the identical action. The same contextual options can be actualized with sound effects, such as thunder or even the cessation of a storm. Choose what you think best tells the story.

The Art in Sound Cues

There are many artistic choices you can make regarding each and every sound cue. As we know, sound exists in time. How do you want to get into the sound? How does it begin? What is it once it does begin? How long do you want it to last? How do you want to get out of it? How does the sound change and evolve over that time?

What are the aftereffects of that sound? Reverberation and echo? Those tails are part of the sound cue. What do you personally experience once the sound ends? Relief? Emptiness? Does the feeling that you personally experienced during that sound linger even after it is gone? Does another feeling follow or take its place? If you feel it, chances are that someone else will share that feeling. Not everyone. Maybe not even most, but someone will.

Sound Cues and Dramatic Style

Like most elements of theatre, the sound cues you create can contain characteristics of one of many dramatic styles. When you create them, it would serve you to choose the theatrical style you will use, or you may decide to experiment with styles that have not been created, or at least styles that have not been cataloged in the common dramatic school of thought.

You might choose a style in which all your sound cues will be consistently created, but you do not have to. It might be more interesting or meaningfully dramatic to stagger the styles in which you create them. This too is a choice. Keep an open mind about what best serves the production.

Furthermore, the style of the sound you choose does not have to comply with the style of other design elements in the same show. The costumes may be realistic, for example, while the delivery of dialogue and action can reflect a heightened realism. Sound cues built in a style that conflicts with other design elements may be a stronger choice. They may harmonize with the style. Conversely, the relationship between the styles of other elements of the play may distort the conflicting elements involved, and that may be a stronger choice. Keep an open mind. Keep all options on the table and decide what tells the best story.

Realism

In realism, sound cues mimic a realistic experience, where the sound delivered to the audience is as close to a real sound that they would hear in the real world, outside of a dramatic production. These can be achieved through recorded medium or live sound (which utilizes the actual thing that makes the sound) or sound effects.

Expressionism

Expressionistic sound seeks to express an emotional experience, rather than to create a realistic sound that can be found in the material world.

Expressionistic sound is almost anti-realistic, exaggerating one truthful characteristic while minimizing other characteristics. Although the sound itself is modified, it can still be recognizable or representational.

Symbolism

Symbolic sounds are aural metaphors that seek to express a truth rather than a realistic sound. A specific sound is represented by a different sound, although the context does not change. For example, a cat's meow might be replaced with a lion's roar in order to make a statement about the cat.

Absurdism

Absurd sound seeks to be unreasonable, inappropriate, or illogical to the conscious mind. It can be manifested by adding a sound that does not fit the context, like an elephant trumpet in New York City, or distorting a sound that does fit in context in a way that is unexpected or illogical.

Surrealism

A surreal sound is bizarre or illogical, while attempting to create meaning in the subconscious mind.

Melodrama

Melodramatic sound cues attempt to evoke an emotional reaction and are usually over the top. The way to achieve melodrama in sound is to pick an emotion you want to express and highlight the qualities of the sound that are traditionally associated with that emotion, while de-emphasizing the other emotional quality of the sound. Melodrama, of course, is often used in parodies since the attempt to manipulate the audience's emotions is so overt that they have the opposite effect.

Incidental, Accidental, and Impromptu Sound

As sound is everything you want the audience to hear, with some circumstances, you might want to build in room for accidental or impromptu sound. For example, the play leaves room for an actor to ad-lib a moment of the play, and they decide to do a little tap dance. You want to be ready to catch the sound of the taps. In fact, you will want to be ready to catch whatever the actor decides to do in the moment.

Representational Sound: The Ubiquitous Woodblock and the Suspension of Disbelief

The woodblock is a workhorse in the world of sound effects. It is kind of a wildcard because it can be used to represent the sound of anything percussive. It has been used extensively in cartoons, movies, and live shows alike. What is it about the woodblock that makes it so universal?

A woodblock is technically a drum, fashioned from a single piece of wood, usually teak. It has a slit cut out of it near the top, the side where the musician strikes it. It can be struck with a stick like a drumstick or a round wooden mallet, usually, but you can strike it with other objects such as your knuckle or a spoon. Experiment with different objects until you get just the sound you want out of it.

It is versatile but often used for a comic feel. Imagine two characters on stage under a tree, and one character is hit in the head with a falling acorn. If you are going for a comic feel, a woodblock might be the perfect sound effect. You can get two woodblocks that create different tones and strike them alternately in rhythm to indicate the ticking of a clock – the passage of time. You can use the two of them as a metronome. Need a sound effect for someone knocking a door with a knocker? Use a woodblock. What about a prowler sneaking around the house? Use a woodblock synchronized to their steps. Throw a stone over a cliff – slide whistle paired with a woodblock at the end of it. Pirates digging for treasure – use a woodblock to indicate the shovel making contact with the treasure chest. You could even use it in sync with an inch worm traveling across the stage, folding and unfolding its body as it progresses. Click, click, click, click.

The woodblock is believable for a few reasons, but the main reason is because of an audience member's willing suspension of disbelief. It is a state of mind where an audience member decides to forget that what they are seeing and hearing onstage is not real while, at the same time, holding onto the ability to abandon that belief when their emotions become uncomfortably intense. It is a safe state to stay in while experiencing things that they might not ordinarily experience – to taste emotions like fear or concern or comedy without becoming inescapably enveloped in them. They can go only so far as they allow themselves to go.

The woodblock is an example of a representational sound effect. The sound of an acorn hitting someone's head is certainly not realistic, but we believe it anyway. Imagine what a real acorn hitting someone sounds like. That sound is dependent on how close you are to the impact event. Typically, the audience is at least several feet if not yards away from the actor. At that distance, the impact would be silent. The audience would certainly not hear it. What about the actor onstage next to the person getting hit? The reality is that a person would have to be pretty close to the victim to hear such a quiet sound. What if you were the victim? You would most likely hear it. You would definitely feel it. The pain you would feel would vary, but it would be real. The emotional response to that pain would be intense, most likely.

The sound effect is from the perspective of the victim – the person who gets hit on the head with an acorn. It represents not so much the sound but the emotional reaction a character experiences as a result of the incident. You are delivering that sound effect to the audience to inform them of that very emotional (painful) reaction experienced by the character. We can sympathetically mimic that emotional response of pain as a function of our human physiology and neurology.

Sound effects are often delivered from the character's experience. For example, I did a show with a fight scene, where a character murders another by twisting the victim's head and breaking her neck. If you were the murderer because you are so close to your victim, you might hear the neck bones breaking, but if you were as far away as the audience would be, you would not hear a thing. We delivered a sound effect from the perspective of the victim. We used lasagna strips hidden in the murderer's glove and amplified it through a wireless microphone hidden on his wrist. Audience members reacted to it with audible gasps. It was effective. The reason it was effective was twofold:

- The sound of lasagna breaking is representative of bones crunching because it involved many small percussive snaps. One snap would not be believable, but several, like 20, all timed in fractions of milliseconds from one another, mimicked the sound of something being crushed.

- The sound was amplified to exaggerate what the victim would hear. The victim would hear it rather loudly, in her ears and in her body, much more loudly than any other person, including the murderer who was very close.
- The exaggerated volume of the sound effect represented the intense emotional response experienced by the victim. As a sound designer, I delivered that representational sound to the audience via a microphone connected to the theatre's sound system.
- Other sounds that are onstage were minimized in order to avoid drawing focus to this intense reaction experienced by the victim.

There is a Goldilocks zone regarding loudness and the suspension of disbelief. If the sound is too soft, the audience will not understand what the character is feeling. If the sound effect is too loud, the audience would not find it plausible that the sound effect is even connected to the event of the bones breaking, and they release their suspension of disbelief. The only thing you can do is listen to it in tech rehearsal and adjust the volume based on your own response to it. You will have to feel it. Make it feel good for yourself, and try not to project how the audience will react. (Easier said than done.) The more authentic it is to you, the higher the chance that the audience will find meaning in it in the form of an emotional response, whatever that response may be. You will get better with practice. Based on years of experience, I can usually predict with good accuracy how effective a sound effect will be, you will, too.

Tool 2: The Delivery System

Peig Sayers is considered among the greatest of Ireland's Seanchaís. The Seanchaís is a tribe of revered storytellers who practiced the oral tradition. Throughout Ireland's history, the Seanchaís were regarded as the custodians of tradition in old Ireland. They were masters of the art of storytelling. Peig was born in 1873 and as a girl lived in the small, remote townland of Vicarstown in Dunquin, County Kerry, Ireland, at the foot of Mount Eagle. In her autobiography "PEIG," she describes her childhood evenings spent by the fireside, listening to stories told by her father, Tomas Sayers, and Old Muiris (a family friend and neighbor). She recounts how storytelling from these elders shaped her early life amid the hardships and poverty of rural Kerry.[1]

Imagine her as a young girl, sitting by the warm fire in the winter, listening to those stories told by her father. Imagine the sound of his voice, and the look on her face as she receives, understands, and creates images in her head. Imagine the sound that he creates with his vocal cords, and how he shapes that sound with his tongue, teeth, mouth, and lips. Imagine the sound traveling via the air from his mouth to her ears. Picture the oscillating air molecules, the transfer of energy from one to the other. The cycle of positive and then negative pressure was gently applied to her young eardrums. The transduction of that kinetic energy to electrical energy, from her ear to her brain. She understands the meaning behind that energy, and she is enraptured in the story.

This ancient, kinetic method is humanity's first and natural delivery system. It was all we had for several millennia. Humans evolved to receive and perceive sound this way.

Human ears are not the only way sound can be received by the body. In the theatre of the deaf, deaf audiences can find meaning in the vibrations in their bones if that music is loud enough. Some deaf audiences are given balloons to hold between their palms, using the balloons to amplify the vibrations in the air so they can feel them on their fingertips.

Cochlear implants are electronic devices that mimic the eardrum and the cochlea with a microphone attached to an array of electrodes that are surgically attached to the cochlear nerve, which transmits the signal to the brain's primary auditory cortex.

We as sound designers will always have to transduce audio signals into kinetic energy for the human mind to derive meaning from them. Even if we could somehow transmit electrical audio signals directly to the cochlear nerve, that signal will have to be analog, which means a digital signal will have to be converted to an analog signal.

In young Peig's case, the signal is simple and natural, from mouth to ear. The air is the delivery system, the original delivery system, but today, it is by no means our only delivery system. Throughout most of the 20th century, theatre has utilized sound systems of various complexities, usually consisting of microphones, amplifiers, and speakers to deliver the sound to the audience. Look at that system for a moment. At its most basic, the kinetic sound signal is converted into electrical energy by the microphones, that signal is increased by amplifying it, and that stronger signal is transmitted to the speakers that transduce that energy back into kinetic energy. The resulting sound is louder going out than it was going in. Sometimes that signal is converted to radio waves and back again along the signal flow.

Back in the day, using technology invented by Bell Telephone, sound systems added functionality such as filters (the precursors to equalizers), noise gates, and compressors. After that, reverb was invented, along with other processors. Later, digital conversion became widely used, and more and more sound system components were built to manipulate a digital signal rather than the traditional analog. This digital conversion led to the widespread use of signal transfer through the adoption of networking and IP protocols.

Yet, even with all of the changes that audio technology has gone through, the basic functions of the sound system remain the same. Signal goes into the system, signal is manipulated by the system, signal goes out of the system, and it is converted to a form that is acceptable to human physiology.

With this perspective, the measure of a sound system changes a bit when contrasted to what people traditionally think of as a sound system. Instead of measuring a system by how efficient it is at manipulating the audio signal, or perhaps how up to date it is with the current technology, or even how expensive it may be at the moment, the true measure is this: Once the delivery system converts the audio signal back into a form that is suitable for human reception, how close is it to the sound designer's intention for how it should sound? I will explain what I mean.

Since the advent of sound systems over the course of the last century, we have developed expectations of what sounds natural when delivered through a delivery system. I personally grew up with a console television set in the living room that had one passive speaker powered by an amplifier embedded in the circuitry of the television itself. By today's standards, it did not sound "good." It was a monophonic (monaural) signal with a stunted dynamic range, usually boosted between 2 and 4 kilohertz to accentuate the human vocal range. Let us face it, it was bad sound. We never complained – we did not even know the difference, as we had nothing to compare it to except record players with ceramic needles and the family telephone. Even still, our family became lost in the story of the television show every time. Why? Because we had a subconscious expectation of what things sounded like through that speaker, and as tinny and distorted as it was, it sounded natural to us. Two characters having a conversation through that speaker sounded nothing like two people having a conversation in the same room we were in, but something in our minds allowed us to fill in the gaps of that dynamic range. It allowed us to forgive the bad sound and interpolate that sound in a way that gave us the same emotional response as if it were live. We learned what to expect, and our expectations were met.

One important reason we were able to get lost in the story despite the bad sound is that it was consistently bad. If it were intermittent – that is – if it switched from very good sound to bad sound back to very good, even once, we would notice it. That inconsistency would draw our attention. Inconsistent sound is what we call "bad" sound because it does not consistently conform to what is expected. When sound is consistently bad, human brains quickly learn to fill in the gaps.

If I were to go back to that old TV set now, having been exposed to the best and most expensive and up-to-date sound systems in the world, I would probably call it bad sound but quickly forget it and get lost in the broadcast. Most people would. That is an important thing to remember when designing your delivery system. The first step to designing your delivery system is to create an intention of what your sound cue should sound like coming out of that specific delivery system and keep it consistent.

The Venue Is Part of Your Delivery System

If you visit the historic Ford's Theatre in Washington, DC, take a moment to look at the back wall of the mezzanine. Carved into the wall itself is a sort of half parabola about 2 feet deep and 3 feet wide and tall enough for someone to stand inside. If you stand inside that notch at the very back of the theatre during a performance, you will hear every word and whisper an actor utters. No sound reinforcement needed. You will probably hear more performance nuance than half of the people sitting in the audience. The shape of the curve takes advantage of all of the sound hitting the sides of that parabola and reflects that sound to one focal point – exactly where you are standing.

Let us look at a different venue. If you ever go to Greece, you must go to the Theatre of Dionysus Eleutherius in Athens. A square stage with wings on each side played to rising seats of stones, where Euripides, Sophocles, and Aristophanes presented their plays to audiences that heard every word. The slope of the seating area was no accident. Engineers learned from observing actors performing in front of audiences seated on the ascending slope of a nearby hill. They observed that the low frequencies of the sound coming from the playing area were filtered out by that slope while the higher frequencies – that of the vocal range of the actors – traveled all the way to the back of the audience. They built the Theatre Dionysus so that the seating benches mimicked the slope of those hills and experimented with how those vocal range frequencies were reflected and directed by the stage walls, especially the back wall and sides. Those frequencies, intentionally directed, traveled all the way to the last rows.

I like to think of the audiences of old, watching a play at Ford's Theatre, or back during the time of the great Dionysia festivals. The noise floor of everyday life was silent compared to what we are exposed to today. They heard the sounds of nature and humans, contrasted to the motors, machines, air conditioners, refrigerators, and car stereos that we are exposed to today. The absence of the noise pollution of today resulted in better hearing, or at least less loss of hearing. If you went back in time, I wonder if, having been raised in the modern world, you would have been able to appreciate all of the sounds those audiences could.

All venues have characteristics of reflection and resonance. Aural reflection refers to how much sound is absorbed by the building and the things that are in it. If the reflectivity is high, that means little sound is absorbed, and the sound energy will bounce back and continue to travel inside the room until the energy is finally depleted through friction itself, if nothing else. Resonance is the capacity for the room and the objects inside the room to vibrate in response to the sound and, in effect, amplify the energy in kind.

Each venue is unique from any other. Some venues, like cathedrals and gymnasiums, have a high capacity for reflecting sound, and some venues, like black box theatres or large shopping malls, have a limited capacity for reflecting sound. Moreover, each venue reflects certain frequencies more efficiently than other frequencies and that, too, varies from venue to venue. Some venues have more resonance than others, and usually that resonance is more intense with some frequencies than others. That quality of resonance varies even from room to room in the same building.

Every venue has some combination of reflection and resonance; therefore, every venue is unique. It serves us as sound designers to consider those room characteristics and use them to our advantage. Get to know the venue. Sit and listen to it. Take measurements with acoustical software. Whistle and clap your hands to hear what comes back to you. Work with the venue's strengths and adapt to the venue's weaknesses. Of course, by strength I mean how well suited it is to your intentions for the show, and a weakness is how little suited it is to your intentions. A good example of working with the venue is how we equalize (eq) microphones. We configure the equalizer (eq) in a way to highlight those frequencies that the venue eats up, and we take away those frequencies that the room reflects. Yes, you can tame a venue and force the shape of every decibel involved, but you will be more successful in your intentions for the show if you work with the room instead, playing to its strengths.

Set your intention, then decide how you're going to accomplish it.

Imagine yourself occupying the space the audience will occupy. Imagine that you are sitting in that seat and experiencing your sound. How do you want them to hear it?

No two audience members are going to have the same experience in a show. Even if you could do that, which is not possible, but even if you could, then you would still have a different response from every single individual. People decide on the meaning; that is, they choose the meaning they are going to draw from a particular moment in the play. The irony is that it is rare when an audience member is aware that they are choosing their response.

What do I mean by no two people are going to have the same experience, though? Picture yourself sitting in a theatre. You are in the center seat of the center row, looking at the stage. It is a proscenium theatre. The theatre mezzanine has 500 seats. Now imagine a person sitting on a chair at the center stage playing a violin. There is no reinforcement, just the sound coming out of the violin itself. The music is amazing. They are a skilled violinist. You are very moved as a result.

What is going on in the room acoustically? The path from the violin to your ears comes at you in a straight line. There is sound passing to the left of you, but it does not hit your ears. There is sound passing to the right of you, but it does not hit your ears, either. The sound that does hit your ears comes at you. There is sound hitting your face, too. There is sound hitting the bones in your body. Imagine you can feel that sound.

Additionally, as you probably know, the sound coming at you is not the only sound that hits your ears. There is music bouncing off the walls at every angle that hits your ears. Like pool balls banking off the rubber, the sound hits every point on the walls at just the right angle, reflecting to your ears. There is also sound bouncing off the back wall of the theatre, but you are sitting in the center of the house, and by the time that reflection makes it to your ears, it has lost so much energy that you are not aware of it and cannot perceive it.

Now imagine yourself in the same seat, 10 rows closer to the stage. Now the violin sounds louder, which changes the ratio between the origin sound and the reflections. It sounds a bit different than when you sat in the center of the theatre.

Now, instead of ten rows forward, you move to the last row. You still get the origin sound, but at a significantly lower energy level – lower volume. Now the ratio between the origin sound and the reflections has changed again, and in terms of reverb, it sounds wetter. Additionally, you are now close enough to hear the reflections of the back wall, so it sounds wetter still due to the time differential of the source and all those reflections.

Same violinist, same venue, the only thing that has changed is where you choose to sit. Three different seats, three different sound qualities. In each seat, the emotional response is a wee bit different. You derive three different meanings from it.

Let us go further. Now imagine yourself sitting under a balcony. Imagine all the sound coming at you from all different angles except the reflections from the ceiling. How does that sound?

Next, imagine yourself sitting on the balcony. How does that sound? Imagine yourself in one of the boxes, close to the stage but on house right or house left. Wherever you sit, you hear something a little differently. There is no getting around that.

You decide to ameliorate the difference by carving out the theatre into zones and using small speakers to cover each zone. Well, if you do that, you will need to reinforce that violin. That means running it through the sound system. To get the same sound throughout the theatre, you are going to have a different eq curve for every zone, since every zone will have different spatial characteristics. So, since you can never subtract sound, (you can only cease creating it), you would not just hear the sound coming out of the speaker, but you will hear a mix of the speaker and the origin sound mixed with the room reflections. The speakers arrive at your ears earlier than the actual violin, so you are going to have to delay the speakers. That delay time will have to vary from zone to zone. And you end up balancing the speakers with the original sound.

You will never get the same sound in all zones. It is going to sound slightly different in every seat, which is okay, since no two seats have the same characteristics. One seat is a little to the left, and one seat is a little to the right, which translates to a little further from the speakers or a little closer to the speakers. Each seat is placed at a slightly different angle from a speaker's axis, a slightly different proximity to the walls, and some seats are covered by a shallow balcony, etc. All of this adds up to a slightly different experience in every seat. You can never change that, but you can sit in each seat (or zone) and ensure that what the sound system is delivering to each address (seat) is similar and, of course, always in accordance with your intentions. Consider volume, tone, clarity, and approximate a similar aural experience the best that you can. You will never give each seat an identical experience, but you can get close. This is another reason not to focus on making the audience feel a certain way, but to allow each individual to respond emotionally as they will.

The Delivery System Is a Tool, Not a Directive

Just like any tool, you want your delivery system to carry out a particular function. Before you can specify the right tool for the job, you must know what the job is. You must decide how you want your sound cues to sound as they are delivered. You must have a clearly defined understanding of the characteristics of the sound that results from the conversion of energy into a form compatible with human physiology. Then you must understand that no two audience members are going to hear the exact same sound – that is, the characteristic of that sound always varies depending on your position to it, whether it is live or coming out of a speaker or other transducer. Next, you must have knowledge of and obtain the tools that exist that will help you deliver the sound as closely to your intentions as possible. Then you must have the knowledge to be able to utilize that transducer as part of a system that can be manipulated to deliver that sound. I will put that in a bullet list for the sake of simplicity.

The Process of Choosing

Chapter 3 explained in detail the philosophy of the choosing process for sound. Let us now add our tools.

The imagining and choosing of sound for a moment in a production happens almost simultaneously with the imagining and choosing of the delivery system. One follows the other. You imagine what the sound cue sounds like, what the specific audience members should experience, and then you imagine the best way to deliver the sound in service to that experiential intention.

1. Make a clear choice about what you want your cues to sound like as they arrive at a particular audience member.
    a. Understanding that no two audience members will experience the sound cues in the exact same way, assemble your audience zones, and set an intention for each audience member zone.
    b. Acquire the knowledge and practical experience to know which tools (speakers, etc.) will serve those sounds as closely to your intentions as possible. Then determine which of those tools you have access to according to your real-world constraints such as time or budget.
    c. Acquire the knowledge and practical experience to plan how to utilize and manipulate that tool as a part of a larger system, and how to manipulate other components such as a player, mixer, or equalizer to obtain the sound as intended at the end of the chain.
2. Make a clear choice.

You are the emperor of sound. You are the creator. You must have a vision of what you want to create, that is, how you want your cues to sound, and then figure out how to realize that vision.

Create a vision for each and every audience member. You do not have to have a separate delivery system for each audience member, but you can have it. You can give each audience member their own private speaker or headset if you choose to. Alternatively, audience members can be split into groups or zones, and you can create a separate delivery system for each group. You will have a separate output for every zone you create.

Acquire Knowledge

As an artist, part of mastering your craft is to learn about all of the technology that is available to you, whether it is old technology or new. Sometimes, the older technology is just what you will need to create the sound you want to create. Sometimes, it will be as ancient as a thunder roll, and sometimes, you will choose the latest invention.

Remember that new technology breeds new choices, just as the electric guitar created new opportunities for a new kind of music. On the other hand, sometimes older technology can be used in ways that are new or novel to you.

It is important to note that your constraints regarding your tools are limited to those tools that are available to you. An understanding of the limitations of the tools available will probably modify your vision. If all you have are full-range speakers, you may have to adjust your intentions of blasting the audience with powerful sub-frequencies.

You might have to come up with creative ways to use your components to get what you want. I once did a show at a theatre in Albany. The sound system was pretty much set (but I still augmented it to my will). The main speakers on this 3/4 thrust stage were set upstage, each behind a flat. It was a mess. All we could hear out of them was muddy low end. The higher frequencies just could not push through those flats. This was a rock musical, and the sound coming through those flats just would not do. To make matters worse, the house engineer was adamant that they were not to be moved, and he was backed by the artistic director and the producer. Moving them was not going to happen. Furthermore, there was no budget to rent speakers outside of the theatre's inventory. I thought about my constraints and opportunities. I decided to use full-range speakers that the theatre had in stock and place them in front of the flats a little higher than the speakers behind the flats. Now, I had the opportunity to use those as the main speakers, but they did not have the low end we needed for this rock musical. I used both sets of speakers. I equalized the full-range speakers to put out only frequencies higher than 200 with a high-pass slope that started around 100 hertz, then I put a low-pass filter on the back speakers with a slope that ended around 250. Then I of course put a proper delay on the forward speakers. I essentially created a disjointed crossover system. By utilizing the gear I had in a creative way, I was able to save money and still get the full higher and lower frequencies at the volumes I needed. The low end was not as clean as I would have liked; shooting through those flats made it sound a bit muddy, but, hey, it was not terrible. Besides, the bass back line was onstage and helped. Sometimes your work is a triumph, and sometimes you have to say, "Hey, it wasn't terrible."

Exercises for Chapter 6

Exercise 1: Sound Cue Design with Intention

Description: Practice setting intentions to guide sound cue creation.

Instructions:

1. Select a scene from a play.
2. Write a statement outlining your intention for a sound cue (e.g., to convey a character's anxiety).
3. Create a sound cue (live or recorded) that reflects this intention.

Exercise 2. Perspective and Imaging Experiment

Description: Explore perspective and image of origin techniques to enhance audience immersion.

Instructions:

1. Choose a theatrical moment (e.g., a character hearing a distant explosion).
2. Design a sound cue from two perspectives: a primary character's peri-personal space and the audience's observer perspective.
3. Submit a description of each cue, explaining how perspective and imaging techniques create distinct experiences. Include a diagram or sketch of speaker placement if necessary.

Exercise 3. Journaling Exercise: Dramatic Style Exploration

Description: Reflect on applying dramatic styles to sound cues while building your daily journaling habit.

Instructions:

1. Journal 500 words each day for one week (seven entries) on how you would use two dramatic styles to design sound cues for a chosen play. Discuss how these styles inform your choices, enhance meaning, and interact with other design elements (e.g., costumes).
2. Submit all entries with a reflection on how daily journaling deepened your understanding of stylistic choices.

Exercise 4. Woodblock Versatility Challenge

Description: Experiment with the woodblock's representational potential and the concept of suspension of disbelief.

Instructions:

1. Create three distinct sound effects using a woodblock.
2. Perform or describe each effect, varying striking objects or techniques.
3. Write an analysis explaining how each effect supports a specific theatrical moment, leverages "Suspension of Disbelief," and aligns with "Context."

Exercise 5. Venue-Driven Delivery System Design

Description: Apply knowledge of venue acoustics and delivery systems to optimize sound delivery.

Instructions:

1. Visit a local theatre space. If you do not have access to a theatre space, a large room will do.
2. Analyze its reflection and resonance characteristics by clapping and listening for the decay of the sound.
3. Write down your thoughts about a delivery system for a sound cue in that space (e.g., an underscore) with specific speaker placements and EQ settings to enhance the venue's strengths. Include a sketch of the speaker layout.

7

The Sound Designer's Technique

Technique refers to specific methods or procedures in a particular creative field, as opposed to a general process. A general process would be painting a landscape on your canvas, while an example of a technique would be practicing the skill of painting a specific mountain ridge on a canvas with a palette knife.

There is an abundance of technologies available to the sound designer, and they are free to choose from any of the techniques employed throughout history. They do not have to stick to the current technology if it does not serve their purposes. By the same token, the sound designer is empowered to make up their own technology if they so choose.

The imagining and choosing of sound for a moment in a production happen almost simultaneously with the imagining and choosing of the delivery system.

> Step 1: You imagine the sound cue. You imagine its characteristics such as length, timbre, volume, source location, and imaging location.
>
> Step 2: Imagine how you would like the sound cue to sound at a particular point in the house or in a particular zone. All audience zones should be considered.
>
> Step 3: Imagine the best way to deliver that sound to that audience member or zone. It might be with speakers, live, or some combination of the two.

In this chapter, we are going to consider the above steps for two specific plays: *A DOLL'S HOUSE* by Henrik Ibsen and *THE TEMPEST* by William Shakespeare. As we do, let us set them both in an imaginary theatre, we will call The Festival Theatre. It is a simple venue with a proscenium stage. There

DOI: 10.4324/9781003519393-9

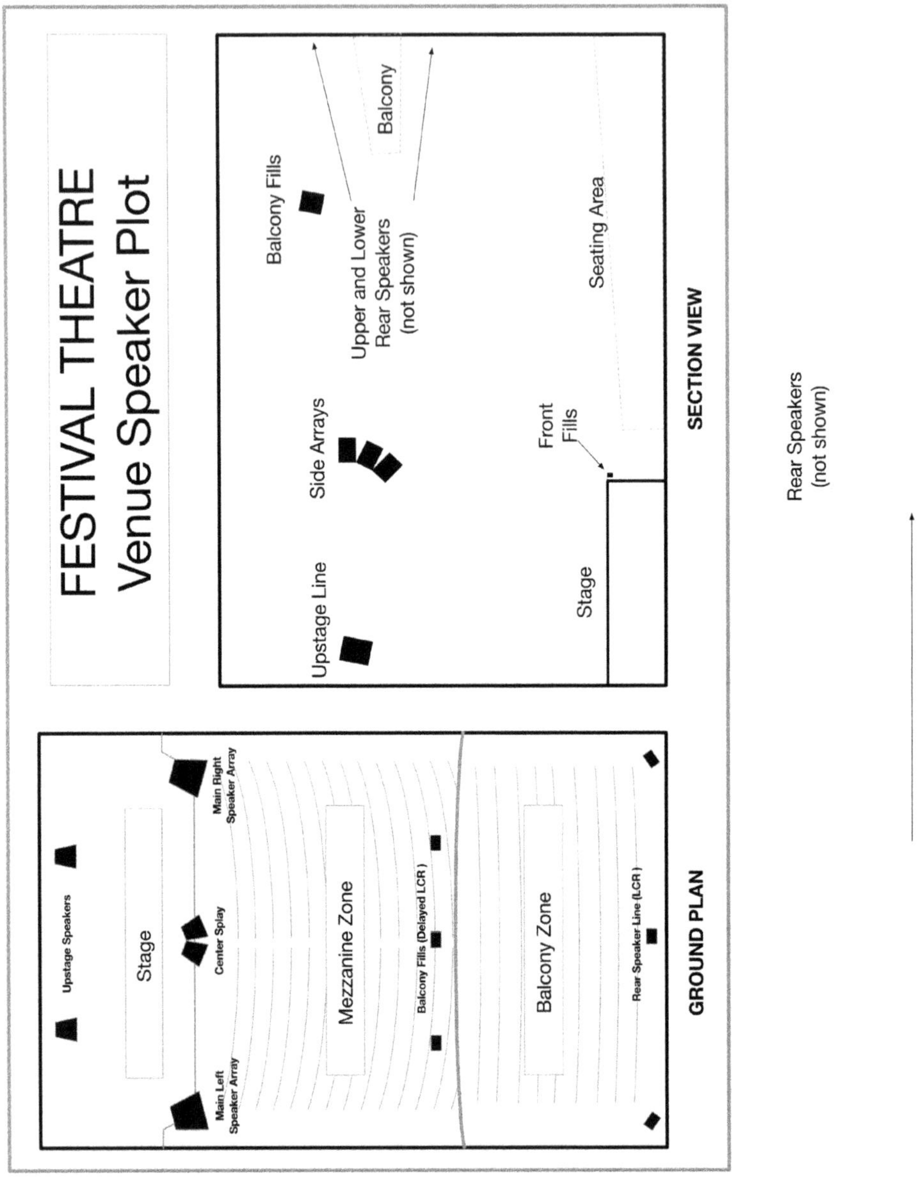

Figure 7.1 The Festival Theatre rep speaker plot.

Drawing by the author.

are two main seating areas: 350 seats are in the mezzanine, and 150 seats are in the balcony. One hundred of those mezzanine seats are placed under the balcony. The control room is at the rear of the theatre on the balcony level, giving the stage manager and lighting technician a bird's-eye view of the stage and audience. We will imagine the sound engineer on the mezzanine level toward the house left. They are in the house, so they can hear the sound as the audience hears it and react if something unexpected comes up.

The Festival Theatre is equipped with a repertory plot or "rep" plot for short. By rep plot, we mean the standard speaker configuration for the space without additions or changes for any specific production. This is a plot that the sound designer starts with when designing their show. Every show will have some changes for the specific needs of a production, for example, an additional speaker may be placed in a specific location to image a sound effect or music in a place the rep plot cannot. Perhaps the production requires a telephone, and you want it to be practical (instead of recorded) and to control when it rings. Perhaps the production requires an additional Wi-Fi router to wirelessly trigger some kind of automation, like a cell phone ringing or a doorbell. In a rep plot, all those additions will be struck after the production, and the venue's speaker plot will be returned to rep.

A Doll's House by Henrik Ibsen

Let us first look at the three steps in action with the opening moments of Ibsen's *A DOLL'S HOUSE*.

Read the opening notes and get an image in your head of the scenery. Ibsen set his play in a house in a Norwegian town circa 1879. A list of sounds (which will become sound cues) in order are:

- A bell rings in the hall (from the entrance hall stage left, unseen).
- A door is heard to open (from the entrance hall stage left, unseen).
- Nora shuts the door (from the entrance hall stage left, unseen).

A preliminary sound plot would look like this:

Table 7.1 Preliminary Sound Plot

CUE NUMBER	DESCRIPTION	IMAGE LOCATION	NOTES
	A bell rings at the door.	Entrance hall, SL	Nora rings the bell, and the maid lets her in.
	The entrance door is opened.	Entrance hall, SL	
	The entrance door is shut.	Entrance hall, SL	Nora shuts the door.

Table by the author.

A DOLLS HOUSE
By Henrik Ibsen

ACT I

[SCENE.—A room furnished comfortably and tastefully, but not extravagantly. At the back, a door to the right leads to the entrance-hall, another to the left leads to Helmer's study. Between the doors stands a piano. In the middle of the left-hand wall is a door, and beyond it a window. Near the window are a round table, arm-chairs and a small sofa. In the right-hand wall, at the farther end, another door; and on the same side, nearer the footlights, a stove, two easy chairs and a rocking-chair; between the stove and the door, a small table. Engravings on the walls; a cabinet with china and other small objects; a small book-case with well-bound books. The floors are carpeted, and a fire burns in the stove. It is winter.

A bell rings in the hall; shortly afterwards the door is heard to open. Enter NORA, humming a tune and in high spirits. She is in outdoor dress and carries a number of parcels; these she lays on the table to the right. She leaves the outer door open after her, and through it is seen a PORTER who is carrying a Christmas Tree and a basket, which he gives to the MAID who has opened the door.]

NORA
Hide the Christmas Tree carefully, Helen. Be sure the children do not see it until this evening, when it is dressed. *[To the PORTER, taking out her purse.]* How much?

PORTER
Sixpence.

NORA
There is a shilling. No, keep the change. *[The PORTER thanks her, and goes out. NORA shuts the door. She is laughing to herself, as she takes off her hat and coat. She takes a packet of macaroons from her pocket and eats one or two; then goes cautiously to her husband's door and listens.]* Yes, he is in. *[Still humming, she goes to the table on the right.]*

HELMER
[calls out from his room]. Is that my little lark twittering out there?

NORA
[busy opening some of the parcels]. Yes, it is!

HELMER
Is it my little squirrel bustling about?

NORA
Yes!

HELMER
When did my squirrel come home?

NORA
Just now. *[Puts the bag of macaroons into her pocket and wipes her mouth.]* Come in here, Torvald, and see what I have bought.

1

Figure 7.2 *A Doll's House* by Henrik Ibsen, Page 1, Project Gutenberg.

Henrik Ibsen, *A Doll's House*, trans. R. Farquharson Sharp (Project Gutenberg, 2008), https://www.gutenberg.org/ebooks/2542.

HELMER
Don't disturb me. *[A little later, he opens the door and looks into the room, pen in hand.]* Bought, did you say? All these things? Has my little spendthrift been wasting money again?

NORA
Yes but, Torvald, this year we really can let ourselves go a little. This is the first Christmas that we have not needed to economize.

HELMER
Still, you know, we can't spend money recklessly.

NORA
Yes, Torvald, we may be a wee bit more reckless now, mayn't we? Just a tiny wee bit! You are going to have a big salary and earn lots and lots of money.

2

Figure 7.3 *A Doll's House* by Henrik Ibsen, Page 2, Project Gutenberg.

Henrik Ibsen, *A Doll's House*, trans. R. Farquharson Sharp (Project Gutenberg, 2008), https://www.gutenberg.org/ebooks/2542.

This is where the play starts proper, but the show event starts when the stage manager opens the doors to the house to let the audience in to obtain their seats. Here, we are going to add some house music to allow the audience to transition into the show. It is a good place to set the palette. Imagine the house before the doors open. It is relatively quiet after all the sound and lighting checks, after the stagehands and production assistants sweep and mop the stage and place the props. Then, after the blackout check, the stage manager yells, "Doors opening!" They open the doors, and the first few audience members enter the house. The usher greets them, looks at their seat number on their ticket, and guides them to their seats. People are talking, laughing, and getting settled in for the show. This is pre-show. The pre-show music is going to start just before the stage manager opens the doors. Here is what the cue plot looks like now.

This production is going to include pre-show announcements. Often this is done live, but in this case, we have recorded the announcements. It completes by itself, so it does not need an out cue.

The opening of the scene is also the opening of the show proper. Like every moment of the show, the lights and sound are going to have to coordinate, even if they are unrelated and not synchronized. Lighting will transition from house lights into the scene lighting, and this usually consists of at least two goals. The first lighting goal is to get the house lights to out, often a slow fade, and the second is to turn on the scene lights, often a fade up of varying fade times. Often, there is a blackout between those two actions, but sometimes there is not. It behooves the sound designer to discuss this with the lighting designer either before technical rehearsals or during tech. This

Table 7.2 Revised Sound Plot with Pre-Show Music Added.

CUE NUMBER	DESCRIPTION	IMAGE LOCATION	NOTES
	Pre-show music	House	
	Fade Pre-show music to out		
	A bell rings at the door.	Entrance hall, SL	Nora rings the bell, and the maid lets her in.
	The entrance door is opened.	Entrance hall, SL	
	The entrance door is shut.	Entrance hall, SL	Nora shuts the door.

Table by the author.

Table 7.3 Added Pre-Show Announcement

CUE NUMBER	DESCRIPTION	IMAGE LOCATION	NOTES
	Pre-show music	House	
	Fade Pre-show music to out		
	Pre-show announcement.		
	A bell rings at the door.	Entrance hall, SL	Nora rings the bell, and the maid lets her in.
	The entrance door is opened.	Entrance hall, SL	
	The entrance door is shut.	Entrance hall, SL	Nora shuts the door.

Table by the author.

discussion will involve the director for the final approval. In this case, the lighting designer will take house to half with the pre-show announcement.

At this point, the house lights are at full, and lighting will get us into a blackout in two stages, utilizing two lighting cues. The first cue will fade the house lighting down to half intensity, and the second cue will take the house lighting from half intensity all the way out. Black out. After the blackout, the stage lights will fade up.

Let us add some transition music to take us into the scene and then fade out at some point once the scene is set. In this case, we can begin the music of the play in the blackout, and let us start to fade it down to out once we hear the doorbell. Let us add those cues to the list.

Numbering Cues

These are all the cues we have so far, and so we can begin to number them. Once numbered, they are no longer simply show sound, but they are officially sound cues. I have heard that some designers sequence their sound cues with letters, but I find that impractical. Numbers allow for more flexibility when you inevitably add cues in between other cues during the rehearsal process. For the same reason, I begin with the number 5 and increment by 5. Some designers will use decimals, as in SQ1, SQ1.1, SQ1.2, etc. I like whole numbers. They make calling the cues by the stage manager more efficient.

You should always remain open to allow the stage manager to re-number the cues, since they deal with many cue lists when calling a production besides

Table 7.4 Added GO Cue Information and Opening Music

CUE NUMBER	GO Cue	DESCRIPTION	IMAGE LOCATION	NOTES
	Doors open	Pre-show music	House	
	House to half	Fade pre-show music to out		
	A beat after house to half completes.	Pre-show announcement	House	House lights to half
	A beat after the announcement completes.	Top of Act 1 Music	House	With fade to black
	Stage lights begin to fade up.	Fade music into onstage speakers		With stage lights up
	Stage lights complete	A bell rings at the door.	Entrance hall, SL	Nora rings the bell, and the maid lets her in.
	Auto-continue	Fade music to out		
	After a beat	The entrance door is opened.	Entrance hall, SL	
	Nora enters the entrance hall.	The entrance door is shut.	Entrance hall, SL	Nora shuts the door.

Table by the author.

sound, such as lighting cues, automation cues, and stage cues. If it makes their life easier to change my numbers, I am all for it.

You may notice that there is an auto-continue after SQ 30. This means that the fade cue starts at the same time as the cue before it, which is when the doorbell is called, but we do not give it a number. If it were an auto-follow cue, it would begin when the previous cue completes. Combining the doorbell cue and the auto-continue fade is considered a cue sequence.

Now, we are into the scene proper. There are a couple of other things in the script that require choices. The first is the fire in the stove. Do we want to hear

Table 7.5 Added Cue Numbers and Page Numbers

CUE NUMBER	PAGE	GO Cue	DESCRIPTION	IMAGE LOCATION	NOTES
5	1	Doors open	Pre-show music	House	
10	1	House to half	Fade pre-show music to out		
15	1	A beat after house to half completes.	Pre-show announcement	House	House lights to half
20	1	A beat after the announcement completes.	Top of Act 1 Music	House	With fade to black
25	1	Stage lights begin to fade up, and house lights begin to fade to out.	Fade music into onstage speakers		With stage lights up
30	1	Stage lights complete.	A bell rings at the door.	Entrance hall, SL	Nora rings the bell, and the maid lets her in.
	1	Auto-continue	Fade music to out		
35	1	After a beat	The entrance door is opened.	Entrance hall, SL	
40	1	Nora enters the entrance hall.	The entrance door is shut.	Entrance hall, SL	Nora shuts the door.

Table by the author.

the fire? Should we have sound coming from the stove? In some cases, we might, but in this case, let us choose not to have the sound of the fire because after a bit, it could become annoying. In this case, we are not going for that feeling.

Another thing that comes to mind is that Nora puts her ear to the door to hear inside her husband's office to see if he is in there working. She does hear him. Do we want to hear something coming from behind the door? In other cases, we might, but in this case, let us rely on the actor to indicate that she does hear him, confirming that he is indeed in his office.

Now that we have plotted the sequence of sounds that will open the show, and our next task is to make choices about how they will sound.

Top of Act 1 Music

This music takes us from pre-show music and announcement and into the show proper. Although the main purpose is to settle the audience in during the transition and set the mood, this is a great opportunity to give the audience some information.

You can indicate the time period by choosing some existing music popular in Norway of the day or perhaps compose music that is in the same style and instrumentation.

You can choose existing music or compose music that is connected to the energy and mood of the characters at the top of the scene, even if it is anachronistic.

You can connect the music to the over-arching themes of the story, which in this case has to do with the limited opportunities that a woman had in a male-dominated world – perhaps a song about/for/by a woman of the time.

Either of the three above choices can work. You might even incorporate all of those qualities in a single piece of music. Can you find or create transition music that contains all those qualities? On the other hand, you might choose to use something more contemporary if you think it would be more meaningful to this contemporary audience.

Also, consider how the music cue ends. You can fade it to out, as indicated on our cue sheet, or you can edit the music in a way that the piece ends about the time stage lights are at full.

Something I might consider here is to image the sound into the house through the main speakers, and then as the stage lights come up, fade the main speakers to place the image location to onstage by utilizing the onstage speakers. Perhaps instead of fading, the music can be edited to end just before we hear the bell. Why? Consider again what happens in the house as the first music begins to play. The audience gets quiet (we hope). Anyone who is still standing finds their seat and sits down (we hope). The house lights are fading to out. This lighting fade and music prompts the audience to focus their attention onto the stage. Stage lights go up. This lighting and sound transition helps the audience to emotionally transition into the world of the play and lets their focus on the real world subside. We see the room onstage, but no one is there. There is a moment before the doorbell when nothing is happening. Music might be just the thing to fill that empty moment to make it more compelling. The music can end just before the bell so that the audience hears the bell in the clear. A side benefit would be that the end of the music could serve as the cue for the bell.

The Doorbell

The doorbell is yet another opportunity to communicate information. We know it is not an electric bell, as they would not have one. That is, unless you choose a dramatic style such as absurdity. It might serve your intentions. In this case, you choose to have a doorbell of the period and culture. What kind of bell was commonly used? A bell on a rope? A twist-key bell? The right bell will be appreciated by someone in the audience who knows about doorbells through the ages, and for those who do not, it will give them a taste of the culture.

The image location is at the door, in the entrance hallway. I would hang a speaker there for the door sounds or use an actual doorbell operated by a stagehand, and place it at the door.

The Entrance Door

What kind of door would be appropriate? A lighter door or a heavy, solid door with a glass window? If you are not sure, you can always ask the scenic designer. They would know. A solid door could reflect their rise to affluence. What does the latch sound like? Does the door creak a little?

Is there sound when the door is open? Sounds that represent the activity on the street could give the audience information about the city in which they live, as well as the period. It also contributes to the illusion that Nora left the door open. It could also indicate the weather. Wind or at least a breeze. Is it raining? The script does not say. What time of year is it? Would there be birds? Any sound you include that is heard through an open door would of course stop when the door is shut.

When Nora shuts the door, how does she do it? Is it a gentle close or a slam? How would that character shut the door at that moment? If you are not sure, you can always consult the director or the actor herself.

Marking Your Script

Your script will be a tool you will reference from the beginning of the process to opening night and beyond. Details quickly add up, and you will need an accurate system to keep them all organized so that no note is neglected. Also, when you give the stage manager your sound cues to put into rehearsal, this annotated script will leave nothing to ambiguity.

Be as prepared as possible for tech so that when changes need to be made, and they will, you can implement those changes quickly to minimize the time people in the room have to wait for you. Here is a sample of the script properly annotated for sound cues.

SQ 5
PRE-SHOW MUSIC

SQ 10
MUSIC FADE TO OUT

A DOLLS HOUSE
By Henrik Ibsen

SQ 15
PRE-SHOW ANNOUNCEMENT

SQ 20
TOP ACT 1

ACT I

SQ 25
FADE MUSIC TO STAGE

[SCENE.—A room furnished comfortably and tastefully, but not extravagantly. At the back, a door to the right leads to the entrance-hall, another to the left leads to Helmer's study. Between the doors stands a piano. In the middle of the left-hand wall is a door, and beyond it a window. Near the window are a round table, arm-chairs and a small sofa. In the right-hand wall, at the farther end, another door; and on the same side, nearer the footlights, a stove, two easy chairs and a rocking-chair; between the stove and the door, a small table. Engravings on the walls; a cabinet with china and other small objects; a small book-case with well-bound books. The floors are carpeted, and a fire burns in the stove. It is winter.

SQ 30
DOORBELL + MUSIC FADE OUT

A bell rings in the hall; shortly afterwards the door is heard to open. Enter NORA, humming a tune and in high spirits. She is in outdoor dress and carries a number of parcels; these she lays on the table to the right. She leaves the outer door open after her, and through it is seen a PORTER who is carrying a Christmas Tree and a basket, which he gives to the MAID who has opened the door.]

SQ 35
DOOR OPENS

NORA
Hide the Christmas Tree carefully, Helen. Be sure the children do not see it until this evening, when it is dressed. *[To the PORTER, taking out her purse.]* How much?

PORTER
Sixpence.

NORA
There is a shilling. No, keep the change. *[The PORTER thanks her, and goes out. NORA shuts the door. She is laughing to herself, as she takes off her hat and coat. She takes a packet of macaroons from her pocket and eats one or two; then goes cautiously to her husband's door and listens.]* Yes, he is in. *[Still humming, she goes to the table on the right.]*

SQ 40
DOOR SHUTS

HELMER
[calls out from his room]. Is that my little lark twittering out there?

NORA
[busy opening some of the parcels]. Yes, it is!

HELMER
Is it my little squirrel bustling about?

NORA
Yes!

HELMER
When did my squirrel come home?

NORA
Just now. *[Puts the bag of macaroons into her pocket and wipes her mouth.]* Come in here, Torvald, and see what I have bought.

1

Figure 7.4 Properly marked script. *A Doll's House* by Henrik Ibsen, Page 1, Project Gutenberg.

Henrik Ibsen, *A Doll's House*, trans. R. Farquharson Sharp (Project Gutenberg, 2008), https://www.gutenberg.org/ebooks/2542.

The Tempest by William Shakespeare

Let us look at a more involved scene and how the three steps can be utilized. The opening scene of *The Tempest*.

The Tempest
By William Shakespeare
Act 1 Scene 1

```
                                   SCENE 1
                   A tempestuous noise of thunder and lightning heard.
                                 Enter a shipmaster and a Boatswain.

                                    MASTER
Boatswain!

                                  BOATSWAIN
Here, master. What cheer?

                                    MASTER
Good, speak to th' mariners. Fall to 't yarely, or we run ourselves
aground. Bestir; bestir!

                                                             He exits.
                                                       Enter Mariners.

                                  BOATSWAIN
Heigh, my hearts! Cheerly, cheerly, my hearts! Yare, yare! Take in the
topsail. Tend to th' Master's whistle. --Blow till thou burst thy
wind, if room enough!

      Enter Alonso, Sebastian, Antonio, Ferdinand, Gonzalo, and others.

                                    ALONSO
Good boatswain, have care. Where's the Master? Play the men.

                                  BOATSWAIN
I pray now, keep below.

                                   ANTONIO
Where is the Master, boatswain?

                                  BOATSWAIN
Do you not hear him? You mar our labor. Keep your cabins. You do
assist the storm.

                                   GONZALO
Nay, good, be patient.

                                  BOATSWAIN
When the sea is. Hence! What cares these roarers for the name of king?
To cabin! Silence! Trouble us not.

                                      1
```

Figure 7.5 *The Tempest* by William Shakespeare, Page 1, Project Gutenberg.
William Shakespeare, *The Tempest* (Project Gutenberg, 2000), https://www.gutenberg.org/ebooks/2235.

GONZALO

Good, yet remember whom thou has aboard.

BOATSWAIN

Non that I more love than myself. You are a councillor; if you can command these elements to silence, and work the peace of the present, we will not hand a rope more. Use your authority. If. You cannot, give thanks you have lived so long and make yourself ready in your cabin for the mischance of the hour, if it so hap. ---Cheerly, good hearts! ---Out of our way, I say!

He exits.

GONZALO

I have great comfort from this fellow. Me-thinks he hath no drowning mark upon him. His complexion is perfect gallows. Stand fast, good Fate, to his haning. Make the rope of his desiny our cable, for our own doth little advantage. If he be not born to be hanged, our case is miserable!

He exits with Alonso, Sebastian, and the other courtiers.
Enter Boatswain.

BOATSWAIN

Down wiht the topmast. Yare! Lower, lower! Bring her to try wi' th' main course.

(a cry within.)

A plague upon this howling! They are louder than the weather or our office.

Enter Sebastian, Antonio, and Gonzalo

Yet again? What do you here? Shall we give o're and drown? Have you a mind to sink?

SEBASTIAN

A pox o' your throat, you bawling, blasphemous, incharitable dog!

BOATSWAIN

Work you, then.

ANTONIO

Hang, cur, hang, you whoreson, insolent noisemaker! We are less afraid to be drowned than thou art.

GONZALO

I'll warrant him for drowning, though the ship were no stronger than a nutshell and as leaky as an unstanched wench.

BOATSWAIN

2

Figure 7.6 *The Tempest* by William Shakespeare, Page 2, Project Gutenberg.

William Shakespeare, *The Tempest* (Project Gutenberg, 2000), https://www.gutenberg.org/ebooks/2235.

Lay her ahold, ahold! Set her two courses. Off to sea again! Lay her off!

Enter more Mariners, wet.

MARINERS

All lost! To prayers, to prayers! All lost!

Mariners exit.

BOATSWAIN

What, must our mouths be cold?

GONZALO

The king and Prince at prayers. Let's assist them, for our case is as theirs.

SEBASTIAN

I am out of patience.

ANTONIO

We are merely cheated of our lives by drunkards. This wide-chopped rascal -- would thou mightst lie drowning the washing of ten tides?

Boatswain exits.

GONZALO

He'll be hanged yet, though every drop of water swear against it and gape the wid'st to glut him.

A confused noise within: "Mercy on us!" -- "We split, we split!" -- "Farewell, my wife and children!" -- "Farewell, brother!" -- "We split, we split, we split!"

ANTONIO

Let's all sink wi' th' King.

SEBASTIAN

Let's take leave of him

He exits with Antonio.

GONZALO

Now would I give a thousand furlongs of sea for an acre of barren ground: long heath, brown furze, anything. The wills above be done, but I would fain die a dry death.

He exits.

3

Figure 7.7 *The Tempest* by William Shakespeare, Page 3, Project Gutenberg.

William Shakespeare, *The Tempest* (Project Gutenberg, 2000), https://www.gutenberg.org/ebooks/2235.

There is a lot going on in this scene. First, we identify relevant information that we know.

- Place and Period: The English characters are on a wooden sailing ship on the high seas in a terrifying storm. 17th century.
- The characters are filled with fear and tension, as indicated by the Boatswain's dialogue. Be aware that the director and actor can choose a different take on the dialogue – maybe the Boatswain loves storms and is giddy or eerily calm. I doubt it, since traditionally this is a terribly frightening life-threatening emergency, and it is written to be so. You will need to communicate with the director about their vision for the scene. For the purposes of this exercise, we will assume that this is a traditional interpretation. People on the ship are afraid.

Before we choose the sounds that we want to deliver, it will be helpful to list the sounds we want the audience to hear into groups by type:

- Actors' voices. Clearly and with enough volume to be intelligible over the sounds of the sea.
- An active and rough ocean sound through the whole scene.
- Various thunder as indicated in the script.
- A storm indicates hard rain and wind.
- The ship itself. Creaking under the pressures of the wind, the ocean, and the rain.
- At one point, the ship splits down the middle. This has got to be a large spectacle sound from the perspective of the characters and probably heightened for dramatic effect.
- Live sound other than vocals. This includes a Boatswain's whistle and an alarm bell.
- Last, but not least, music will be used to get us into it and to underscore the whole scene. We will also use music to get us out of the scene and transition into the next scene.

Sound Groupings

There is a lot of sound all happening at once. It is our job to keep tight control of every sound in every moment of the scene, and to do that, we are going to organize the sound into groups:

- Dialogue delivered through the public address (PA) system. The actors are going to wear wireless lavaliere microphones.

- Momentary sound cues. These will include sound cues that happen at specific moments, like thunder. These cues will complete by themselves or we will have to stop them with an out cue.
- "Storm at Sea" soundscape. A soundscape is a group of cues that fire together to represent the environmental conditions of the scene. It serves us to think of this as one cue, even though it includes many sound effects playing at the same time. This includes the ocean, the wind, the rain, and the creaking of the ship. Grouping these all together into a separate mix allows us to dip the overall volume at times when other events happen, such as thunder or the moment the ship breaks in two. Dipping the entire mix will add clarity and focus to those momentary cues.
- Underscore. Like the soundscape, music plays continuously from the top of the scene into the transition to scene 2.

We do have the option of throwing all the sound at the audience from the main left and right speakers and simply mix and pan, but if different sounds were imaged from different locations, it would give the scene a more three-dimensional atmosphere. In effect, we are going to mix all the different sound cues in the room from different speaker plot locations, instead of mixing them electronically in the mixing console and delivering them through only two. Doing so will allow us to represent the perspective of the actors more realistically, which would allow the audience to more easily suspend their disbelief of the spectacle. It will also add clarity to each sound, as our brains can more easily make sense of simultaneous sounds from different locations. Let us make our delivery plans group by group. Once all the plans are in place, then we can go back and listen to everything in symphony with each other and adjust the mix for maximum dramatic effect.

Actor Voices

Our first duty with actor voices is intelligibility. We want the dialogue and other vocalizations to be clearly understood. This is our goal for every production, whether it is a play, a musical, a dance show, a musical concert, or any other type of production.

For this production, reinforcing the actors' voices will be a necessity. There is no getting around it. In a smaller venue, you may not need to reinforce the actors' voices at all. It would depend on the spatial relationship of the audience seating to the stage. Do the voices reach the last row intelligibly? It would also depend on other sounds competing with the actors' voices. Even with microphones, I would advise the director to have them all but yell, which conveniently works for the scene, since the characters are panicked.

For this scene, the wind, rain, and roaring sea would probably destroy intelligibility even in a smaller venue.

In any play, deciding not to reinforce the actors may be surprising because these days audiences are almost always exposed to louder-than-life amplification, but it is inevitably a choice open to you. Artistically, it may be more meaningful to an audience to hear the dialogue acoustically. Amplification is an added filter in the signal flow between the voice and the ear. Ironically, the signal must be equalized, processed, and electronically mixed in order to sound natural. If you are not forced to use microphones for practical reasons, choosing not to use microphones for aesthetic reasons is always an available option.

However, for this production of this play in this venue, let us decide to use microphones. What does that delivery system look like? For this, again, we will take perspective into consideration, specifically the perspective of the actors onstage. Since each actor can clearly understand themselves, so should the audience. To do that, we are going to start with a distributed system, dividing the audience into coverage zones and hanging speakers to cover each zone. We hang them for the same reason we moved to the front row to listen to our violinist – to maximize the volume differential between the source and the room reflection and resonance. It also maximizes the volume differential between the source vocals and other sounds in the room. Third, it maximizes the spatial differential between the vocals and everything mentioned earlier.

We are also going to throw the vocals through the main left and right speakers, as well as the center cluster. That means those coverage speakers are going to have to be individually delayed appropriately, so there is no resulting echo.

Rough Ocean

The sound of raging sea waters is terrible, and it will justify the rocking and creaking of the boat. It also justifies the panic that the crew and passengers are feeling. Even more so if we provide the sound of waves splashing onto the deck of the ship periodically. This is an emergency.

From the perspective of the characters, the churning ocean is close. When you are threatened, your brain can focus keenly on the sound of that threat. That is what we are going for. The churning ocean is close and beneath them, and the waves that splash onto the bow are closer and on the same plane. I would start by putting them in the onstage speakers, the main speakers, very slightly in the zone speakers, and in the rear speakers above and below the balcony. When the wave splashes onto the deck, I would raise the volume a bit in all of the speakers, but especially on the zone speakers.

Wind

The wind justifies the raging sea, the ship creaks, and the ship's bell. It is on the same plane as the characters and above them. It is both distant and close. The thing about wind is that it sounds different when it is close than when it is farther away. In this case, I would start by creating several sound cues to have different wind sounds in different speakers. That gives the wind a sense of three dimensions. It also creates the illusion of movement. It simply sounds more realistic.

The following wind sound in my design will ultimately be played simultaneously:

- A legato low-frequency wind played through a pair of left and right speakers. I would put a low-pass filter to eliminate the mids and the highs.
- A midrange wind with filters to eliminate the highs and the lows.
- A wispy high-frequency wind, with a high pass to eliminate low- and midrange frequencies.

To make the wind even more three-dimensional, I would create movement by programming synchronized pans. Take the low-frequency wind, which is a stereo file. The left channel would move from the main left speaker to the rear right speaker, while the right channel moves from the main right speaker to the rear right speaker. That simulates the air travel from the front of the house to the back of the house.

The high, wispy wind is also a stereo file. We can move the left channel from, say, the zone speakers to the main left, while the right channel moves from the main right to the onstage right speaker.

All this movement would go on for the duration of the scene. As you can see, this method can get very complex very fast, and we have not yet dealt with the mid-frequency wind. Paperwork here will be crucial to keep track of what is happening at every moment of the sequence. The payoff with this method is huge. Compare all of that with simply playing wind through the left and right main speakers.

Rain

Rain is similar to wind cues. I would give the rain three dimensions by splitting them up into three cues that run simultaneously.

- Low-end rain with a low-pass filter.
- Midrange rain with filters on both ends.
- High range with a high-pass filter.

I would not program movement in the rain, but I would place them in speakers in pairs of left and right and then keep them there. I would start with the low-end rain in the main left and right, midrange in the rears left and right along with onstage left and right, and high-end rain on the side speakers and maybe zone speakers.

Listen to What We Have So Far

At this point, I would play the wind, the rain, and the sea in the house and listen to how it all sounds mixed together. It will inevitably sound muddy and confusing. This is the time to clear it up with an equalizer. Create holes in the dynamic spectrum of the wind where the water and rain can play through. Create holes in the rain EQ where the water and wind can come through. Do the same with the water. Those holes will not sound like comb filtering because the holes in the dynamic range of one element will be filled by another element.

Along with eq filtering, you will find clarity by adjusting the volumes of each cue in each moment – mixing them in the air, as it were. This is a precise set of tweaks, so take your time with it and use your ears as your guide. As you put it all together, you are going to listen to it over and over again.

Thunder

Thunder has movement as well, but it is usually not right up upon you. It usually begins up above you somewhere, and then echoes in the atmosphere in all directions. Then that echo trails off in one direction. When you create thunder, you want that movement as well.

To deliver thunder, as a start, I would place the initial crack (or boom) as high as I could and directly over the audience. If possible, I would place a left and right speaker focused face down from the ceiling of the theatre, close to the center of the audience. That is where I would image the initial crack. It would represent the location of the thunder as directly above – we are right in the middle of the storm. That sound would echo in the side speakers, rear speakers, main speakers, and finally taper out into one set of speakers, perhaps onstage or rear of house.

Then I would listen to all sounds in the house again, and tweak the mix and timing so that all sounds are represented clearly. In the case of thunder, it is okay if you do not hear the other sounds because thunder, having lots of energy in the full dynamic range of human hearing, will naturally drown out the other sounds. In fact, you might help it along by programming a dip in the other sounds, at least for the duration of the initial thunder crack. Shape that dip in a way that keeps you from noticing it.

The Ship Itself

I would want to hear the wooden planks creaking under the pressure and perhaps a warning bell clanging with the sways of the bow. I would image all of this onstage and perhaps sweeten it in the mains a little for clarity.

Live Sound Other than Vocals

We want to hear the Boatswain's whistle. It wants to be close, as close as the dialogue. I will use the lavaliere microphone that is mounted on the actor to pick up the whistle and deliver it through mains and zone speakers to image it on the ship. There is also a live bell ringing the alarm. This will also be reinforced, and come through the speakers onstage, again, to image it on the ship.

Music

So far, we have talked about sound effects, but music will play a key role in conveying what the actors are feeling. Music will transition into the scene, play underneath the entire scene, and transition out of the scene to resolve into scene 2. Just like the sound effects, it will be recorded. In this case, I draw a relationship between dialogue and music, similar to the relationship between dialogue and blocking/action. Music, like dialogue, is close to the actors because it is an aural representation of how they are feeling. Therefore, I want to deliver it to the audience through the main speakers and zone speakers. As with dialogue, no music will come out of the top or side speakers, nor will it be played through the rears.

Putting It All Together

All of the sound cues above are played simultaneously, and it has the potential to become messy real fast. You will have to build it with care, incrementally, and that will take a bit of time to get it right. Built properly, this has the potential to be a spectacular scene, each moment orchestrated with precision. I would rather try and fail than not try. Besides, failure is not the end of the story. Failure is the middle of the story. The story continues when you learn from what failed and try again until you get it just right.

Paperwork

Now that we have an idea of all the sound elements and how to deliver them, we can start to build our sound plot. As usual, we will go through the script moment by moment.

Pre-Show Sound

As people walk in, I would start with the sound of a period wooden ship at sea, with the ocean and the creaking of the planks of the ship, and a gentle wind that pushes it. That is me. I might consider that as the entire pre-show sound, except that water noises for an extended period of time tend to make people have to go to the bathroom, so music appropriate to the show will be added. We can play it all through the main speakers, which will give a two-dimensional image. That way, there is an emotional heightening when we ramp up from the pre-show sound to the three-dimensional sound we have in mind for the start of the show. The ship-at-sea sound can continue even after we fade the music out and play the under pre-show announcement. Once the announcement is complete, Act 1 music begins as we fade lights to blackout. Then, in the blackout, we can create a sudden crash of thunder and all the sounds for the "storm-at-sea." Basically, thunder happens, and then, it all hits the fan, and we are immediately thrust into a rough storm on a wooden boat at sea.

The music for the scene must be eerie and have a soft touch. It has to sweeten the rest of the sound without getting in the way of the sound. I personally find it hard to talk about music and sound in the specific. What is eerie to me may be a lullaby to you. We must let the sound and music speak for themselves.

Table 7.6 *The Tempest* Preliminary Cue Plot

CUE	PAGE	GO CUE	DESCRIPTION	IMAGE LOCATION	NOTES
	1	A beat before house opens	Pre-show sound	Main speakers	
	1		Fade sound to out		
	1	When fade sound completes	Pre-show announcement	Main speakers	
	1	As light fade	Act 1 Music	Main speakers Zone speakers Onstage speakers	
	1	In blackout	Thunder and storm at sea soundscape		

Table by the author.

A good rule of thumb for the sound plot is to list the cues that will be called by the stage manager or operator and maybe some relevant auto-follow cues for clarity. You may have noticed that some of the cues on the plot are sequences that combine many cues. The show control programming will look much different. Here is what a show control spreadsheet would look like for the thunder and "storm at sea" sound cue. Keep in mind that this is a spreadsheet for only one cue, we will number it 25 for the sake of the example, and the underscore is already playing under all of this.

Table 7.7 *The Tempest* Cue 25 Sequence Plot

CUE	Description	Group	Sound Files	IMAGE LOCATION
25	Thunder and "storm at sea" soundscape	1 – thunder	- thunder	- Ceiling cluster and rears (delay) and sides (delay)
		2 – ocean	- ocean low end - ocean high end - ocean splashing	- subs and mains and zones - sides and rears - sides, rears, mains, onstage
		3 – wind	- Wind low end - Wind mids - Wind high end - Wind wispy	- mains and subs - onstage - rears - onstage and sides
		4 – rain	- rain hits water low - rain hits water high - rain hits ship deck	- onstage and subs - mains and sides - mains and sides and rears
		5 – ship	- ship creak	- mains and sides and rears

Table by the author.

That is the plan for SQ 25. It is the starting point for when you go into the theatre for quiet time to listen, program, and tweak. A few notes:

- The music is composed to roughly match the action and rises in intensity through the scene and crescendos at the end of the scene. Timing can be accomplished in rehearsal. Give the music to the stage manager and director early in the rehearsal process so that they can rehearse with it, and then the music can be edited to match the action once the action has been blocked.
- All of the cues fire simultaneously and play on top of each other. You are definitely going to have a quiet time in the theatre so you can accurately adjust the volume in each zone. That includes volume mix, delay settings, eq settings, etc. This is a very complex scene for sound, and it can get away from you during tech. Ensure that each cue is programmed precisely beforehand. Also, labeling everything descriptively will be crucial to you during tech.
- You will probably have to make sub-mix volume adjustments wholesale during tech once you can hear the vocal microphones while the sound effects play. Often, tech rehearsals will be your first opportunity to hear both. Until then, make your best guess. You will get better at predicting that over time.

Here is the sound plot for the entire scene:

Table 7.8 *The Tempest* Scene 1 and Scene 2 Sound Plot

CUE	PAGE	GO CUE	DESCRIPTION	IMAGE LOCATION	NOTES
	1	A beat before house opens	Pre-show sound	Main speakers	
	1	At show time	Fade sound to out		
	1	When fade sound completes	Pre-show announcement	Main speakers	
	1	As light fade	Act 1 Music	Main speakers Zone speakers Onstage speakers	

CUE	PAGE	GO CUE	DESCRIPTION	IMAGE LOCATION	NOTES
	1	In blackout	Thunder crash	Top speaker Side speakers Rear speakers	
		Auto-continue	Storm at sea sound	All zones	
	1	Master: "Bestir; Bestir!"	Thunder crash	Top speaker Side speakers Rear speakers	
	1	Boatswain:" . . . If room enough!"	Thunder roll	Top speaker Side speakers Rear speakers	
	2	Gonzalo: " . . . our case is miserable!"	Thunder crash	Top speaker Side speakers Rear speakers	
	2	Boatswain: " . . . lay her off!"	Thunder roll	Top speaker Side speakers Rear speakers	
	3	Gonzalo: " . . . against it and gape	Ship splits	Top speaker Side speakers onstage speakers	
	3	Gonzalo: " . . . I would fain die a dry death."	Transition to Sc. 2 – thunder	Main speakers Top speaker Side speakers Rear speakers	
		Auto-continue	Fade music and sound to out		
		Auto-continue	Island soundscape	All speaker zones	

Table by the author.

Here is what the script looks like, marked.

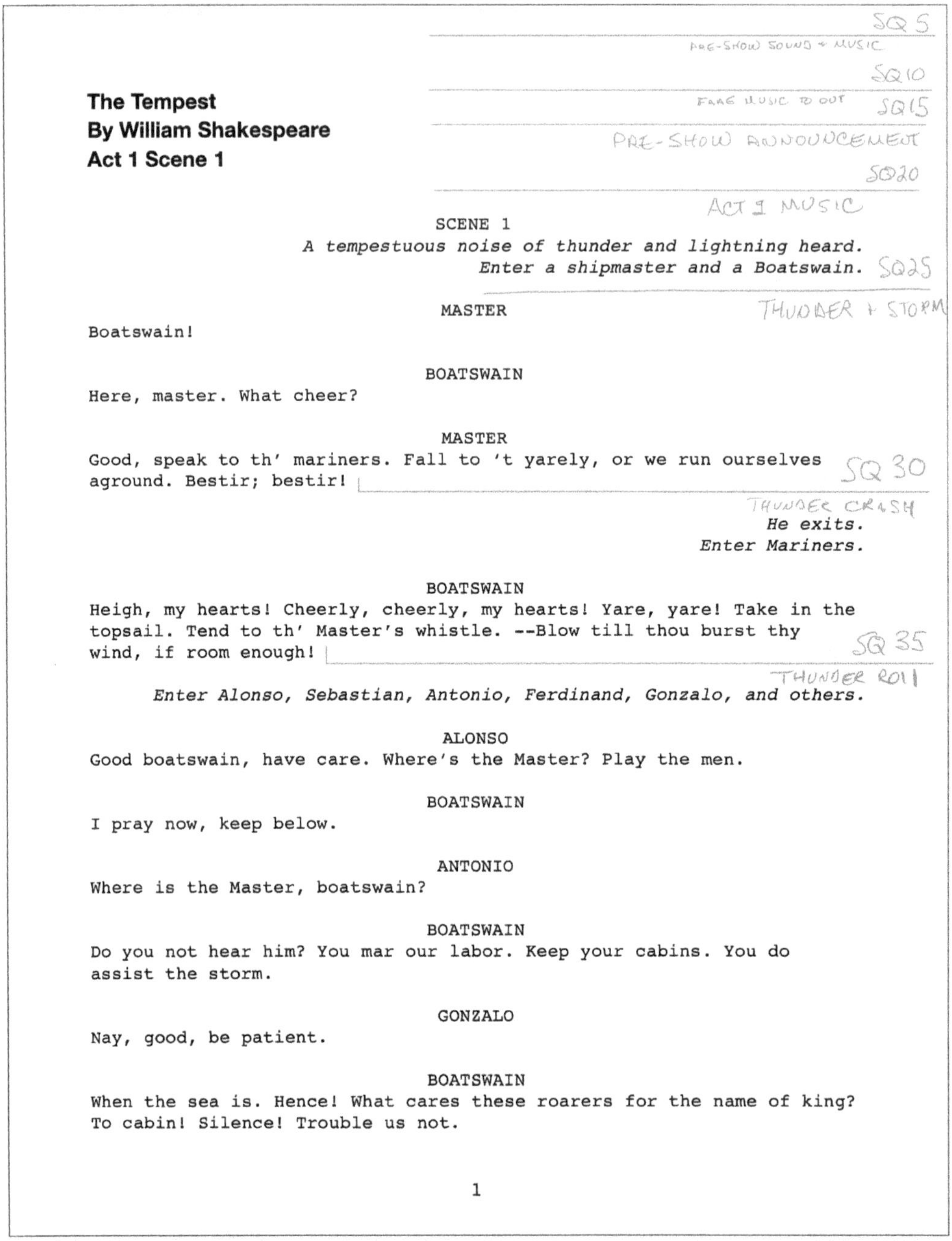

SQ 5
PRE-SHOW SOUND + MUSIC
SQ 10
FADE MUSIC TO OUT SQ 15
PRE-SHOW ANNOUNCEMENT
SQ 20
ACT 1 MUSIC

The Tempest
By William Shakespeare
Act 1 Scene 1

SCENE 1

A tempestuous noise of thunder and lightning heard.
Enter a shipmaster and a Boatswain. SQ 25
THUNDER + STORM

MASTER

Boatswain!

BOATSWAIN

Here, master. What cheer?

MASTER

Good, speak to th' mariners. Fall to 't yarely, or we run ourselves aground. Bestir; bestir! SQ 30
THUNDER CRASH

He exits.
Enter Mariners.

BOATSWAIN

Heigh, my hearts! Cheerly, cheerly, my hearts! Yare, yare! Take in the topsail. Tend to th' Master's whistle. --Blow till thou burst thy wind, if room enough! SQ 35
THUNDER ROLL

Enter Alonso, Sebastian, Antonio, Ferdinand, Gonzalo, and others.

ALONSO

Good boatswain, have care. Where's the Master? Play the men.

BOATSWAIN

I pray now, keep below.

ANTONIO

Where is the Master, boatswain?

BOATSWAIN

Do you not hear him? You mar our labor. Keep your cabins. You do assist the storm.

GONZALO

Nay, good, be patient.

BOATSWAIN

When the sea is. Hence! What cares these roarers for the name of king? To cabin! Silence! Trouble us not.

1

Figure 7.8 Marked script, *The Tempest* by William Shakespeare, Page 1, Project Gutenberg.

William Shakespeare, *The Tempest* (Project Gutenberg, 2000), https://www.gutenberg.org/ebooks/2235.

GONZALO
Good, yet remember whom thou has aboard.

BOATSWAIN
Non that I more love than myself. You are a councillor; if you can command these elements to silence, and work the peace of the present, we will not hand a rope more. Use your authority. If. You cannot, give thanks you have lived so long and make yourself ready in your cabin for the mischance of the hour, if it so hap. ---Cheerly, good hearts! ---Out of our way, I say!

He exits.

GONZALO
I have great comfort from this fellow. Me-thinks he hath no drowning mark upon him. His complexion is perfect gallows. Stand fast, good Fate, to his haning. Make the rope of his desiny our cable, for our own doth little advantage. If he be not born to be hanged, our case is miserable!

SQ40
THUNDER CRASH

He exits with Alonso, Sebastian, and the other courtiers.
Enter Boatswain.

BOATSWAIN
Down wiht the topmast. Yare! Lower, lower! Bring her to try wi' th' main course.

(a cry within.)

A plague upon this howling! They are louder than the weather or our office.

Enter Sebastian, Antonio, and Gonzalo

Yet again? What do you here? Shall we give o're and drown? Have you a mind to sink?

SEBASTIAN
A pox o' your throat, you bawling, blasphemous, incharitable dog!

BOATSWAIN
Work you, then.

ANTONIO
Hang, cur, hang, you whoreson, insolent noisemaker! We are less afraid to be drowned than thou art.

GONZALO
I'll warrant him for drowning, though the ship were no stronger than a nutshell and as leaky as an unstanched wench.

BOATSWAIN

2

Figure 7.9 Marked script, *The Tempest* by William Shakespeare, Page 2, Project Gutenberg.
William Shakespeare, *The Tempest* (Project Gutenberg, 2000), https://www.gutenberg.org/ebooks/2235.

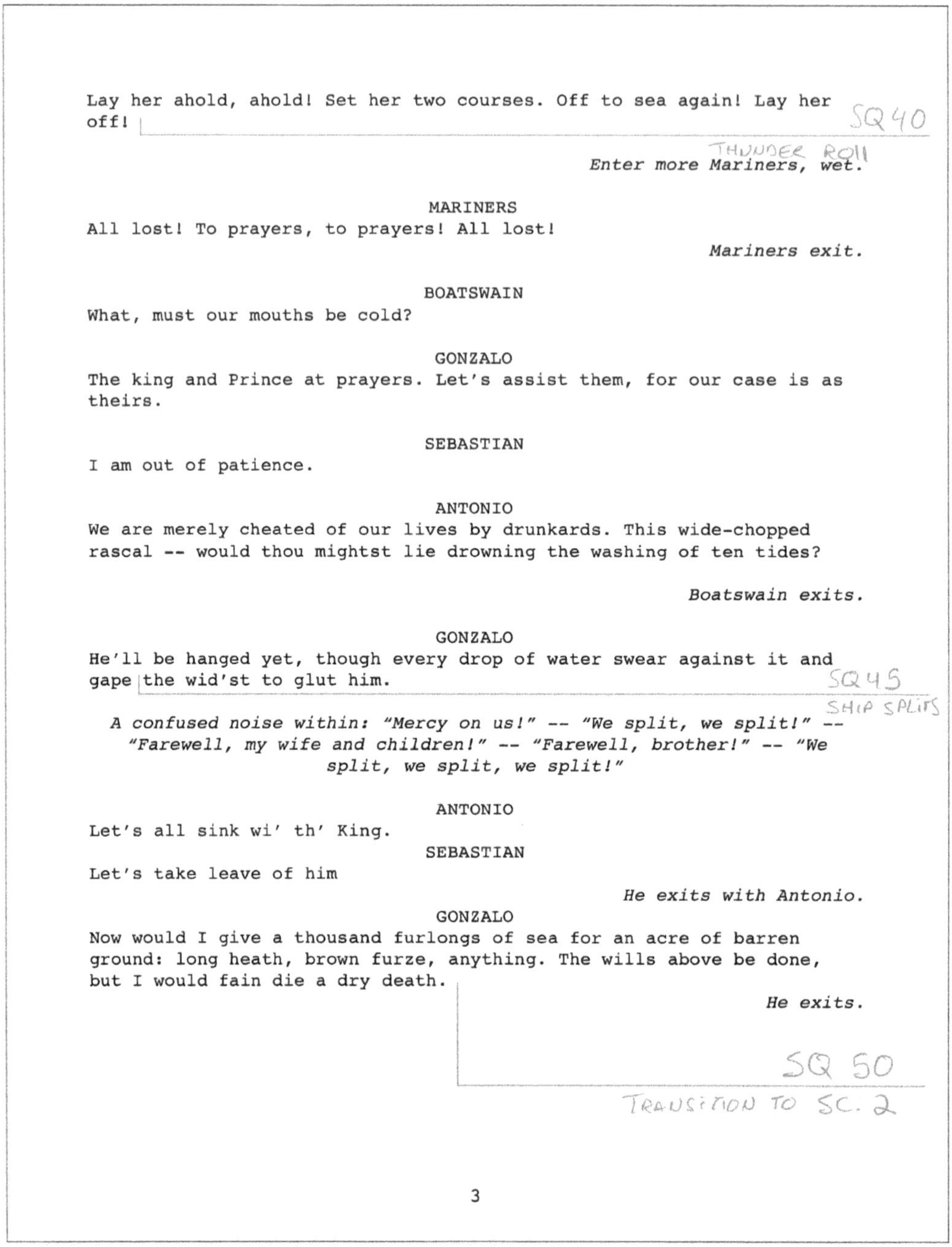

Lay her ahold, ahold! Set her two courses. Off to sea again! Lay her off!

SQ40

THUNDER ROLL

Enter more Mariners, wet.

MARINERS

All lost! To prayers, to prayers! All lost!

Mariners exit.

BOATSWAIN

What, must our mouths be cold?

GONZALO

The king and Prince at prayers. Let's assist them, for our case is as theirs.

SEBASTIAN

I am out of patience.

ANTONIO

We are merely cheated of our lives by drunkards. This wide-chopped rascal -- would thou mightst lie drowning the washing of ten tides?

Boatswain exits.

GONZALO

He'll be hanged yet, though every drop of water swear against it and gape the wid'st to glut him.

SQ45

SHIP SPLITS

A confused noise within: "Mercy on us!" -- "We split, we split!" -- "Farewell, my wife and children!" -- "Farewell, brother!" -- "We split, we split, we split!"

ANTONIO

Let's all sink wi' th' King.

SEBASTIAN

Let's take leave of him

He exits with Antonio.

GONZALO

Now would I give a thousand furlongs of sea for an acre of barren ground: long heath, brown furze, anything. The wills above be done, but I would fain die a dry death.

He exits.

SQ 50

TRANSITION TO SC. 2

3

Figure 7.10 Marked script, *The Tempest* by William Shakespeare, Page 3, Project Gutenberg.

William Shakespeare, *The Tempest* (Project Gutenberg, 2000), https://www.gutenberg.org/ebooks/2235.

Transition to Scene 2

It would be my intention that the music crescendos at the end of this scene. The ship is fractured, and the characters are panicked. It looks like many will lose their lives. There is thunder to help that feeling. The next scene takes place on the safety of the island after a short passage of time. Sound can represent that passage of time and help us transition to that feeling of relief when the characters reach dry land by simply crossfading the storm at sea soundscape into a more serene island soundscape, with island birds and water lapping on the beach.

Crossfades have two events to them. One sound fades to out while a new sound starts. Sometimes that new sound starts immediately at full volume, and sometimes, it will fade up to volume from a lower volume or from silence. This crossfade eases us through the emotional and environmental transition.

Develop Your Own Technique

This chapter has described the tools I would use and how I would use them for this production in this house with these actors and this director at this time. These are my choices, and that is the point. If you are the sound designer, it is up to you to make the choices you find meaningful. It is also up to you to master your technique by learning from others and from your own experience. Make it your goal to use them in your own way. Experiment. Bring the best of your own creativity and skill toward your goal of creating meaning in each and every moment of the story. Remind yourself, "No one has put on this play at this time in this venue with this cast and this sound designer." Create it as if it were never done before because it *has* never been done before under these circumstances.

Spatial Audio

We designed *The Tempest* with different sounds coming from different locations. There are several ways to accomplish that. We did it here manually, programing each location and fading granularly. Another way to do that is to use the technology of spatial audio, which allows you to encode the audio in a way that is also "three dimensional." You can consider it another tool in your toolbox. You can choose to use it, as with all tools, when it serves your purposes conceptually and fits within your external and self-imposed constraints. That is the point. You are not bound to use any specific technology. Use any and all technology that delivers sound to the audience in a way you intend.

Prepare for Tech

This is as good a time as any to talk about preparing for tech rehearsals. Time is money, as they say, and I personally do not ever want to waste everyone's time waiting for me while I work on something I could have prepared beforehand. Here are some standards I personally practice that may work for you:

- Ensure all sound cues have been rehearsed with. I like to submit a first draft of the sound cues at the first rehearsal so that they can be tweaked and adjusted for timing as much as possible by the time we move into tech. Ideally, the director has heard these cues before the first rehearsal, but that is not always possible. Understand that the cues would not sound like they will in the theatre venue, but they will be close. Remember your intentions and be ready to adjust those intentions based on what you learn in rehearsals.
- Ensure all cues are programmed and ready to go at the start of the first tech rehearsal. This allows you to adjust during tech based on what you learn in the space, instead of wasting time trying to catch up. On the same note, make sure all your gear is installed and working properly before you ever go into tech. During tech, I like to put my full attention on the specific moment we are working on and not troubleshooting things that have not been tested and are not working.
- Ensure all networking is trouble-free, and all the networked gear is talking to each other.
- Scope out your workspace. Ideally, it is a tech table equipped with all the gear you require to monitor every part of the delivery system. That includes your show control software, mixing console, wireless microphone systems, etc.

These are only a few examples. The idea is that I want to ensure all systems are working, all cues are firing, and not one thing is left undone. I want to be free to be as flexible as possible during tech so that my team and I can serve the show to the best of our ability

Exercises for Chapter 7

Exercise 1: Sound Cue Plot Development

Description: Practice creating a detailed sound cue plot, using the examples from *A Doll's House* and *The Tempest*.

Instructions:

1. Select a short scene from a play.
2. Identify at least five sound cues you would like to include in the scene.
3. Create a sound cue plot that includes cue numbers, descriptions, image locations, and notes.
4. Submit the plot and an explanation of how each cue supports the scene's narrative.

Exercise 2: Three-Dimensional Soundscape Design

Description: Apply techniques like panning and time differentials to create a three-dimensional soundscape.

Instructions:

1. Design a 45-second soundscape for a storm scene in a hypothetical play, incorporating three elements (e.g., wind, rain, thunder).
2. Use at least two speakers to simulate movement via panning or delays.
3. Submit a recording of the soundscape, along with an analysis explaining your technique choices and speaker placement.

Exercise 3: Cue Sequence and Crossfade Experiment

Description: Practice creating a cue sequence with a crossfade to manage complex transitions.

Instructions:

1. For a scene transition (e.g., from a chaotic to a calm setting), create a cue sequence with at least three sound cues, including a crossfade (e.g., storm to serene island soundscape).
2. Record the sequence, ensuring one cue auto-continues.
3. Submit an explanation of your choices, how the crossfade enhances the transition, and how you used *Crossfade* and *Cue Sequence* to maintain narrative flow.

8

The Sound Designer's Process

While your technique is the way you use the tools available to you, your process is how you go about creating a complete piece from start to finish. You use tools to shape the moment. You have a process to shape the whole and bring it to fruition.

You are going to have to discover your own individual process. No one can teach you how to navigate the creation of your art. It is like a toddler learning to walk. You really cannot teach a child to walk; they must figure it out for themselves. You can only encourage them. They test a step and find out if they did it right, or they fall. It is nature. They do not pout about falling. They may cry when their hind quarters slam against the floor. Would not you? Yet the failure itself does not faze them. It is all data and the exuberance of earnest practice.

In this book, I present theory; the ideas and the concepts and how to think about sound design, but your process is not theory, it is practice. It is not conceptual, but it is execution. To be meaningful, it must be authentic. To be authentic, it must come from your individuality. The only way to learn how to do it for yourself is to do it for yourself. Do not let anyone get in the way.

With all that in mind, however you navigate the business of designing a show, in whatever order you use the tools, what you do is going to fall into one of the five stages of design, and it will serve you to understand these stages of design.

While your process is as unique to you as signing your name, you sign your name using letters of the alphabet. While your process is as individual

DOI: 10.4324/9781003519393-10

as your signature, everything you do and every tool you use as part of your process will fall into one of these stages:

- Planning.
- Communicating.
- Building.
- Execution.
- Archiving.

I am not numbering these stages. They are not sequential, and none of them depends on the completion of another. Some things you do will fit in more than one stage. The goal is to design and create a completed production. How you get there is up to you.

I have found that my process differs from show to show. Sometimes the difference is slight, while some shows require a radically different way of working. The process can be messy. It is often uncertain. Every process includes certain goals that you will have no idea how to accomplish. Have confidence that once you get into the studio and begin working, you will come up with a creative solution. That is the value of big dreams; they often force you to rise to meet them. Confidence is when you do not know how you are going to accomplish something, but you trust yourself that when you get there, you will get it done.

Collaboration

So far in this book, collaboration has been mentioned repeatedly. It should happen in every stage of your process. You work with others in service to the production. No one collaborator is more important than the show itself. As a sound designer, you will need the cooperation of the costume department, the lighting department, the music department, the scenic department, the props department, and all the other departments. You will also need the cooperation of the actors. You will need the cooperation of everybody, and they will need yours. Creating a solid production requires a joint effort of all departments toward a common goal.

Collaboration is one of those words that seem to always be on the lips of theatre makers. It ranks right up there with words like "community" and "storytelling," as if the person uttering the word is reminding himself as much as who he is talking to. The harmonious partnership of ideas requires the boldness to communicate your ideas while balancing an open-mindedness to the ideas of others. It is an art more than a science, and we in the technical fields would do well to master it. When you collaborate with the right people toward creating a cohesive and compelling show, it can be quite fun and even joyful, and everyone ends up creating better work.

At its most basic, collaboration is the act of working with someone to produce or create something. When you think about all the people involved in mounting a theatre production, it would be impossible not to succeed without working with and in concert with the parties concerned.

Any production starts with the story creator. Normally that would be the playwright, and in the case of a musical, the composer (or in the case of Stephen Sondheim, composer/lyricist). They come up with the story and the dialogue and communicate it through the script. As the sound designer on a project, we collaborate with the playwright's script: Analyzing it for themes and story arc, and mining information from script notes. New plays usually consist of much collaboration between the playwright and the director.

The director plots the show's heading by making countless choices regarding themes, tone, style, etc. The director's ideas for how the play will look and feel is what we mean when we use the term "Director's Vision." The director will then assemble their designers and communicate the vision, and the designers will create based on his choices. The designers will also collaborate with each other, discussing how they think moments will look and sound based on the director's intentions. Then the actors arrive, and rehearsals begin, and the director communicates blocking, the action, the way of the characters, and allows the actors to bring the best of their technique to the production. The best directors bring out the best in the actors. Meanwhile, the designers are creating and building the show, communicating their work and their progress with everyone. During technical rehearsals, everyone gathers into the venue and works together to bring all the ideas and plans that were collaborated on into fruition. We continue to work together through dress runs, previews, opening night, and sometimes all the way through closing.

Collaborating is always in service of the show and the director's vision. Yet, just because the director drives the main impulses of the production, that does not mean we should wait for the director to tell us exactly what to do. We have our own ideas and suggestions on how best to tell the story that the director and playwright want to tell. The dynamic between the director and designers (and everyone else, for that matter) is one of a leader and sometimes shepherd who brings on experts to create an aspect of the show according to their specialty. At best, it is a harmonious partnership of ideas that lead to something greater than either could imagine on their own.

Theatre collaboration is the harmonious interplay of the shared visions of skilled artists concentrating on the particular role they play in the creation of a production.

The Planning Stage

All plans exist in your mind. Paperwork and drawings are only memory aides for you. Paperwork is also the way in which you communicate your design to others during the planning and building stages. I am including electronic documents in what I call paperwork.

Often, the choices you make will happen simultaneously with the paperwork and drawings you create. You decide on a sound you want to hear, and then you add it to your sound plot, then you decide how to deliver it to the audience, and then you annotate that information on the sound plot. You draw each section of your signal flow diagram as you decide how to deliver each sound cue to the audience. You take the ground plan of the venue and the set and annotate where the speakers will be placed as you draw your signal flow diagrams and system diagrams. Then you go back to your sound plot and look at the next sound you have specified. It is not mystic, but it is holistic, and how you do it is entirely up to you.

Gathering Information

- To plan your show, you are going to need some information about it, and the more information, the better the result. You will need to understand the story that the playwright, the director, the design team, and all the other collaborators are going to tell.
- Know the story the team is going to tell.
- Read the script.
- Make a list of all the sounds listed in the script.
  - Make a list of all transitions in the script where you might want to add sound.
  - The top and end of acts are good places for sound.
  - Scene changes are good places for sound.
  - Location changes are good places for sound.
  - The passage of time is a good place for sound.
- List all the different locations in the script – do you want soundscapes for some of them?
- Meet with the director to get their take and discuss yours.

- Listen to the director before you speak. What does the director have to say about the play and the story they want to tell?
- Collaborate with other designers.
  - What is the rest of the design team thinking? Set designer, lighting designer, costume designer, or anyone else you will be collaborating with?
- Create or obtain a cast list.
- Become familiar with the venue.
  - How big is the stage area?
  - How large is the seating area? What is the seating position in relation to the stage?
  - Analyze the room for reflectivity and resonance. Get to know the characteristics of the room.
  - Are there any noises involved with the venue, like heating, ventilation, and air conditioning (HVAC) noises, or creaky seating?
  - Can you hear anything from the outside environment?
  - Identify all of the idiosyncrasies of the space that create noise.
  - What noise can you eliminate and what noise can you not eliminate?
- How long will the show run? How many performances?

More Choices to Make

Next, you will make choices based on the information about the production you have gathered. Most of the time, you will not know everything about the production from the start. Decisions such as casting, blocking, and rehearsal schedules will be made when the time comes. Make the decisions when you can and avoid locking yourself into decisions before you know enough about them.

Visualize the show playing out in your imagination. Listen to potential sounds you may hear. Imagine the actors speaking the dialogue. Jot down notes.

I like to start with reading the script. I have known very accomplished, award-winning designers whose first exposure to the show is the designer run – when the actors run the show in rehearsal with the designers attending. I also like to look at it again while sitting in the venue itself, if possible. Even if it is a venue I am intimately familiar with, it often inspires me to sit in the middle of the room and imagine the action taking place on the stage. If I do not have the opportunity to sit in the venue, I do it in the studio. I start with the first moment, which is the house doors opening and the audience coming in and settling into their seats, and then I imagine the next moment. I imagine one moment to the next, through the first half, through intermission, through the second half, all the way through to the audience walking out, and the doors closing.

Making Plans

- Create a sound cue plot.
- Mark your script.
- Decide how to deliver each sound cue.
- Create an input list.
  - Make a list of recorded cues.
  - Make a list of actor microphones.
  - Make a list of orchestra microphones.
  - Other live microphones, such as off-stage singers or voice-overs.
  - Stage manager's voice of god (VOG) microphone.
  - Any other input sources.
- Create a list and diagram of output zones.
  - Speakers for the audience.
  - Monitor speakers for the actors.
  - Monitor speakers for the orchestra.
  - Monitor speakers for offstage areas.
- Create a list and diagram of subordinate systems you are going to use.
- Create a list of all devices that sound will control automatically (also known as show control).
  - Projections.
  - Lighting cues.
  - Live MIDI instruments.
  - All other device sounds may need to be controlled.
- Create a signal flow diagram.
  - Which console channels will accommodate which input sources.
  - Make it complete. Include all snakes, cables, connectors, processors, amplifiers, and speakers.

Rehearsals

Rehearsals are part of the planning process. Build sound cues early enough so that they can be put into rehearsals so that the actors and director can work with them. Attend rehearsals. Is there an opportunity to improve your sound cues in some way? Are there other opportunities to add sound that could only become apparent during rehearsals? This is your time to experiment with sound.

The Communicating Stage

It is hard to talk about your sound before you create it. That is because sound communicates things that words cannot adequately express. Sound communicates things that exist below words – ideas that are more fundamental than words. Yet, even though words will never accurately describe the sounds in your design, it behooves you to be as communicative as you can during every step of your process.

Create Your Sound Cues Early

The best way to do this is to create your sound cues as early as you can, at least rough drafts. In fact, consider every iteration, update, and tweak a rough draft, subject to change. There are many advantages to creating your cues early in the production process:

- It is easier and more accurate to communicate your intentions to the director and designers if they have something to listen to, as opposed to a verbal description, which is neither easy nor accurate.
- Getting your cues into rehearsal early on allows the actors and director the opportunity to work with and react to your cues. They can adjust their timing to the cue. It also allows the director the opportunity to give the sound designer notes on how to tweak the cue, or whether they even want a sound cue at that moment. It also gives the stage manager that much more time to learn when to call the cue. Although, to be fair, I have never met a stage manager worth their salt that has ever had any significant trouble calling a sound cue, even when placed at the last minute.
- Creating your sound cues early also gives you the chance to create your show control programming early, which saves time in load-in and tech.

Diagrams and Other Paperwork

Your sound plot and marked script communicate to the stage manager exactly when and where to call the cues. Make sure you get that to the production stage manager as soon as you have that information. If there are any changes or additions, get that to the stage manager quickly as well.

System drawings and gear lists communicate your design to the engineers who will be building the P.A. systems. Include all systems, including distributed program systems to the front of house and green rooms, as well as party line communication systems such as Clear-Com systems.

Do not take anything for granted in your system diagrams. Include every cable and connector in the signal flow. Do not assume that anyone will know what you intend for those details, and do not make them guess.

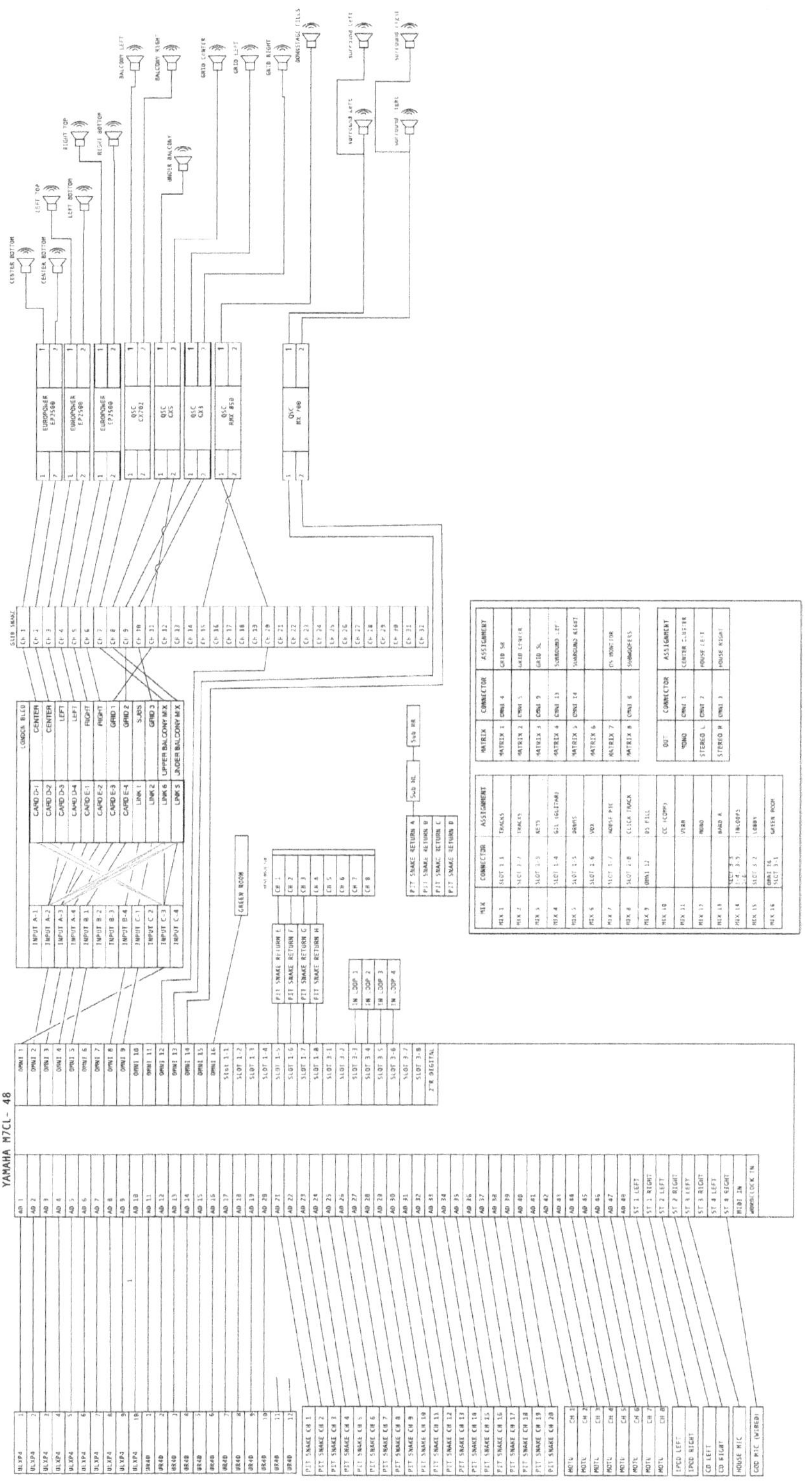

Figure 8.1 Signal flow diagram for mainstage, circa 2015.
Drawing by the author.

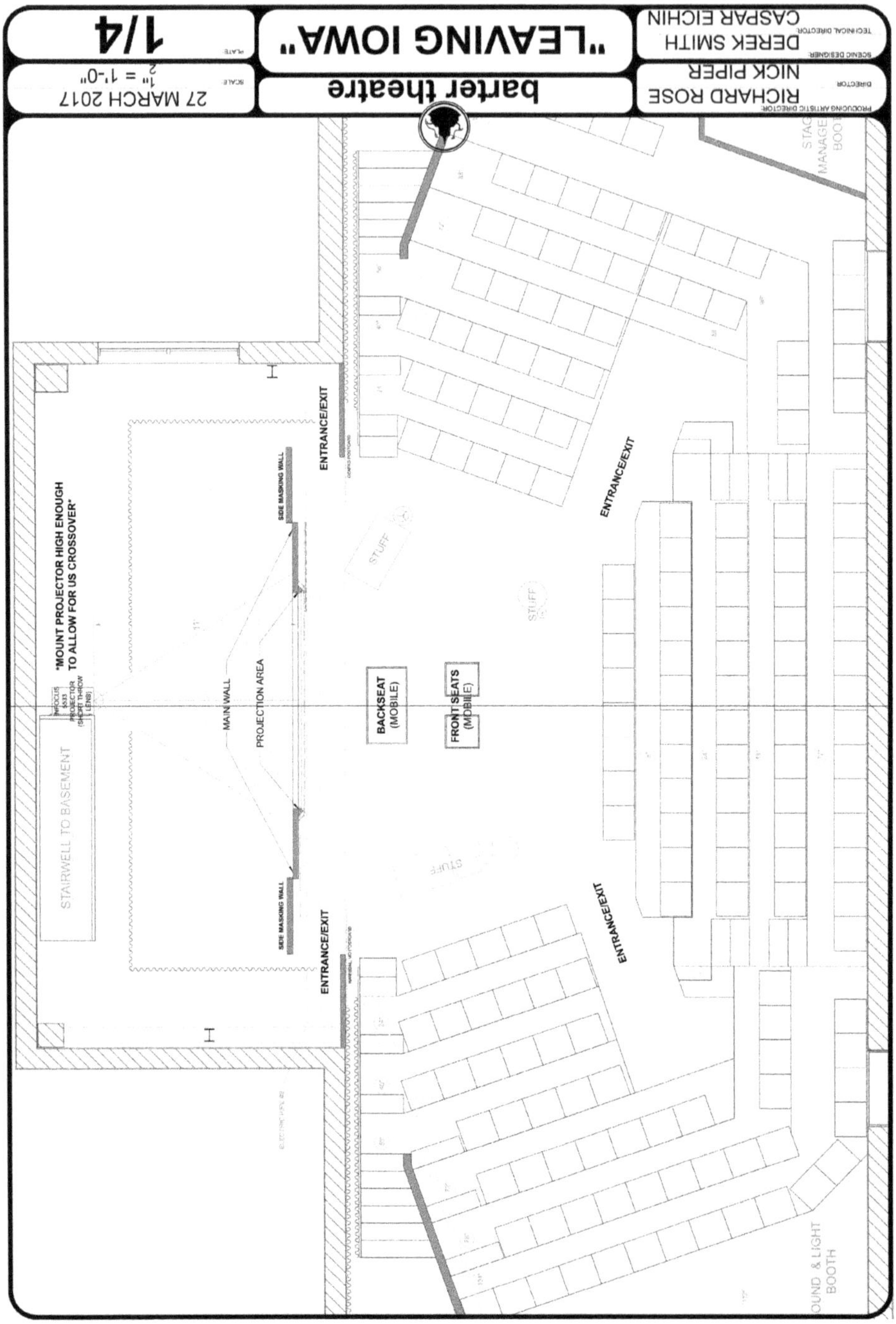

Figure 8.2 Ground Plan, circa 2016. Here, you will draw your speaker locations, racks, and other equipment. Drawing by scenic designer Derick Smith.

The Building Stage

In this stage, you will compose music, create sound effects, assemble the delivery systems, and anything else you want to create in order to execute sound. Anything you do to eliminate sound fits into the building stage.

To-Do List for the Building Stage

- Record and edit sound cues, including sound effects, music, and composed music.
- Program the show control computer.
- Build the PA system.
- Build the show control system.
- To control any automation that might be controlled by the sound department.
- To control the playback of sound effects.
- To control sound system automation.

The Execution Stage

Execution of your sound happens anytime your sounds are used in a run of the show, of a scene, or even just a moment. Rehearsals are the best chance to listen to your sound with actors and, if applicable, music. Try to be in rehearsal whenever new sound cues are being used. It will provide you with data you cannot get anywhere else. The data you are looking for here include things like actor timing, relative volume, and how aurally the timbre of the cues mixes with the timbre of the actors' voices. Is the sound cue too fast or slow? It gives you a chance to discover how to make the sound cue fit better. After rehearsal, you can go back and build a different sound cue. Then you put it into rehearsal again and listen to it played in rehearsal. Expect that cycle to continue until you (and the director) are pleased.

Your sounds are also executed in technical rehearsals. You get to listen to your sound under show conditions. That means you can hear the cues with actor voices, music, in the venue, on the stage, and the house with scenery, costumes, and everything else. This is another opportunity to find where and how you might make your cues better and more meaningful.

The same applies to previews, where you can listen to the show under show conditions, with the added element of a live, reactive audience.

Of course, the execution stage also includes the run of the show. Once opening night happens, you might still be tempted to make changes but avoid that temptation unless there is something so wrong that not fixing it would be irresponsible.

Someone once said that shows are never finished, only abandoned.

Archiving

Archiving is not really part of the design, but it is a good habit to get into as part of your process. Back up all your work and keep it for future reference. You might want to keep backups of each iteration of your work. This serves two purposes. First, it gives you a body of work from which you can draw. You can take the best of it and create a portfolio to give to producers or other hiring agents. It also gives you a record of the show in case it is ever remounted. This happens now and again, and you never know which show will be picked for a remount. It also serves as a repository for sound effects that you have created should you ever want to use one in a future show.

You should also keep a copy of the finalized show, electronic paperwork, and all programming you did. You will want to keep it forever. If nothing else, it will be good content for your hall of fame exhibit. Who knows, they may also serve as good teaching tools should you ever want to teach.

In the future, your archives will be considered artifacts of your time. They will serve as another piece of the puzzle of what life was like when you were alive. What people were like, what was considered style and art. Your archives will tell another story and inspire meaning once again to people you will never meet. Maybe even future AI. Who knows?

Exercises for Chapter 8

Exercise 1: Communication Strategy Development

Description: Enhance skills in the communicating stage by preparing materials to share with a production team.

Instructions

1. For a hypothetical production, draft a sound plot for the stage manager and director.
2. Create a marked script page showing cue placements.
3. Submit the sound plot, script, and a reflection on how early communication supports collaboration.

Exercise 2: Journaling Exercise: Process Reflection

Description: Reflect on your personal sound design process across the five stages, inspired by Chapter 8's emphasis on individuality and practice (page 1), while building a daily journaling habit.

Instructions:

1. Journal 500 words each day for one week (7 entries) on how you would navigate the planning, communicating, building, execution, and archiving stages for a chosen play. Discuss challenges, creative solutions, and how your process reflects your authentic voice.
2. Submit all entries with a reflection on how daily journaling clarified your process.

Exercise 3: Building Stage Technical Setup

Description: Apply the building stage by assembling components for a sound design.

Instructions:

1. Design a delivery system for a sound cue (e.g., an underscore) in a 500-seat venue, specifying equipment (e.g., microphones, speakers, console) and show control programming.
2. Record and describe one sound effect you would create.
3. Submit a report detailing your setup and programming.
4. Submit a sketch of the system layout.

Part 2

A Brief History of Sound Design

Sound design has been practiced since the beginning, even if they did not have a word for it. Humanity usually does not recognize the existence of a thing until they have a word for it. Storytelling, from the earliest days, requires that the story be heard by listeners. Add to that the natural human tendency to make sounds with their voice to heighten the impact of the story, and sound effects are born.

Imagine a tribe of Palasgians in Greece around 7000 BC sitting around the campfire after a successful hunt. Bringing back an aurochs (the ancestor of our cow) was a big deal and cause for celebration. (Our most joyful religious celebrations are called feast days for a reason.) On rare occasions, the hunters would bring back a large brown bear, whose bones, hides, fat, and meat meant warmth, shelter, light, and, of course, a feast. The tribe would dance in celebration, chant with joy to their gods, and tell the stories of the hunt.

As the crowd sits, the leader of the successful hunting party stands in front of the fire and the tribe, and he begins to tell the tale of how the brave hunters approached the ferocious bear, stalked it, and eventually felled it. The entire tribe sits in complete silence and listens, captivated by the story. He mimics the bear's roar, and the crowd gasps. He mimics thunder from the heavens, and the crowd jumps. He mimics the wail of the wounded bear, and the crowd cheers.

Then, just in the middle of the story, a voice from the back of the listeners calls out, "I can't hear you! Speak louder!" Chances are, the people in the back had a hard time seeing the storyteller as well, as they had to look through the large group of listeners. Inevitably, an older member of the tribe whispers to his friend seated next to him, "What did he say? What did he say?"

DOI: 10.4324/9781003519393-11

From that point on, it has been a focus of theatre-makers to ensure that the audience understands the story they are telling. The effort to enhance the meaning that the audience derives by employing sound effects such as vocal mimicry began at the same time.

In Part 2 of this book, we will cover select periods in Western history that were significant in their advancement of the theatre tradition. As we do, we will see that those two goals, intelligibility and the inspiration of meaning through sound effects and music, existed from the beginning of time and endure even today. We will explore how theatre-makers went about achieving these goals. The effort shaped architecture, it shaped the advent of scenery, it collaborated with costumes, it shaped actors' vocal and physical techniques, and an entire music genre was invented to ensure that the actors' lines and lyrics were easily followed. Elaborate machines were invented and employed to create sound effects to draw the audience in, and, ultimately, create meaning with their sound.

Ethel Caterham is the oldest human in the world as I write this. She was born in the village of Shipton Bellinger, Hampshire, England, on August 21, 1909. That makes her almost 116 years old. She was 18 years old in 1927 when she took a job as a nanny for a British family in India. In 1933, she married an officer in the British Army at Salisbury Cathedral. After a long and happy marriage, her husband passed away in 1976. She drove his Triumph Dolomite until she was 97. She has outlived both of her daughters. In 2020, she contracted COVID-19 and survived. So far in her lifetime, she was a subject of King Edward VII, King George V, King Edward VIII, King George VI, Queen Elizabeth II, and King Charles III. On April 30, 2025, she became the oldest living person in the world.

Ethel grew up in a time when the noise floor of life was quite low compared to today. While I can remember a time when there were no cell phones or computers (or even debit cards), Ethel can remember a time before sound reinforcement systems, and in Shipton Bellinger, recording and playback devices pretty much did not exist. She can remember a time when music was always live, and stories were told on the front porch, or in church, or at the pub. It was an oral tradition. She lived through the evolution of sound technology that has changed humanity forever, and so she has a unique perspective that we cannot have. I never lived in a time without telephones or television, or recording and playback devices like the phonograph. Have you? We can only imagine her perspective, and doing so would do us good.

When we think about sound effects for the theatre, we think about microphones and recorded playback. That is the way it is done these days. You might say it is our contemporary tradition. Our recording and reinforcement tradition is wonderful because it makes creating sound more accessible to people who are not acoustical scholars, it enables the fast and easy manipulation of sound and opens doors to creating all kinds of sound that have never existed before. We have brand-new ways to deliver them to audiences everywhere.

The question I pose is, why should you? Does current technology create more meaning for the audience than did the ancient ways? You have the right to choose to forego designing in the way everyone else does. You can choose to create sound employing any technology from history if it better serves your intentions. You can even do things no one has ever done before. Who is to say one tradition is better than another? You are not required to comply with your contemporaries. You do not have to use recorded sound. Certainly, do not use recorded sound simply because everyone else is doing it.

I ask again, who's to say which tradition is better? The answer is you are. If you can create more meaning using older techniques, then give yourself permission to do so. You are the sound designer. You are the artist. You are the sound designer artist.

9

Early and Late Classical Period – The Greeks and the Romans

Western theatre was pioneered by the Greeks and, later, the ancient Romans. Hellenistic rituals and Dionysian festivals evolved into the structured dramatic forms we still use today.

The ancient Greeks viewed sound, music, and choral performances as a way to honor Dionysus and to promote a collective civic identity. They were both religious ritual ceremonies performed in front of large audiences, which presented the great stories of their gods and their myths to them. Dionysus, the god of wine, fertility, and celebration, was honored with choral songs and dances, such as stories of satyrs, the half-human, half-animal companions to Dionysus. Dithyrambs, tragic hymnic choral performances, were part of the festival. These ritual songs and dances evolved into more formal dramas that included tragedy and comedy. Playwrights such as Aeschylus, Sophocles, and Euripides wrote plays that relied heavily on dialogue as part of these religious and civic rituals.[1]

The playwrights and performers clearly intended that the audience hear the sacred stories and myths to spiritually connect with divine narratives and to also understand and participate in their cultural practices and traditions. Greek venues could accommodate thousands of audience members. The audience needed to understand the lyrics of the songs presented, and they needed to understand the complex dialogue and poetry of the drama, even if they were sitting as far away as the back row. Greek practices emphasized the inspiration of meaning through deliberate choices

DOI: 10.4324/9781003519393-12

shaped by clear intentions, context, and constraints. In Greek and Roman theatre:

- The intention was to ensure audibility and emotional resonance through vocal intelligibility and propagation of sound effects to inspire emotional impact.
- The context included large amphitheaters, like the Theatre of Dionysus in Athens (seating 12,000–15,000) and the Theatre of Pompey in Rome (up to 20,000).
- The constraints involved the size of the venues and the need to overcome ambient noise such as wind.

The goal of vocal intelligibility and emotional inspiration has endured throughout the history of Western theatre, and they remain our goals as sound designers today. It serves us to examine how they went about accomplishing these goals. By exploring vocal techniques, acoustic architecture, masks, sound effect technologies, and the contributions of key figures, we can see how sound was used to create meaningful experiences.

Greek Sound Priorities

Greek theatre evolved from musical and choral performances at Dionysian festivals into complex dramas that relied heavily on plot and dialogue. They recognized how important it was for audience members to understand the sacred stories. They established two critical sound-related goals that remain our goals today:

1. **Intelligibility of Actors' Vocals:** In the open-air theatres that were used in this era, actors needed to project dialogue clearly to massive audiences.[2] Clarity was essential for conveying intricate narratives. They relied on vocal training, venue acoustics, and textual structures to ensure every word was understood, as we shall explore.
2. **Sound Effects to Enhance Audience Experience**: The Greeks peppered rudimentary sound effects to amplify the emotional and ritualistic impact of performances, particularly during festivals. They used machines to simulate thunder, and there is unconfirmed evidence that they had a wind machine. These effects engaged audiences emotionally, reinforcing the narrative and cultural significance of the performance.

The Romans inherited these Greek traditions and adopted and expanded on the methods to achieve these goals. They performed in larger venues, such as the Theatre of Pompey, which could seat 18,000 to 27,000 attendees. They introduced advanced technologies to create dramatic soundscapes. Both cultures used sound to create meaning, intending that audiences not only heard but felt the stories.

Vocal Intelligibility

As dialogue became more important to dramatic programs, the Greeks and later the Romans, driven by necessity, cleverly pursued vocal intelligibility. They applied a multipronged effort.

- Actors practiced a combination of vocal techniques.
- Acoustical theories informed building design.
- Masks were worn that resonated in the vocal range.
- Choral recitation and other textual structures were used, with the intention of maximizing clarity and audience comprehension.

Vocal Techniques for Intelligibility

Actors trained rigorously to project their voices and sustain that projection through repeated performances of lengthy plays and musical programs. They practiced several techniques to make sure every theatergoer heard their performance.

One essential technique was vocal training with breath control. They practiced extensive breathing techniques to project with force and resonance across the theatre for long periods of time.[3]

Another important method was to use stylized diction and articulation. Deliberate enunciation and elongated vowels helped with the clarity of their words and elevated delivery, especially for highly poetic dialogue.[4]

The rhythmic and musical delivery of dialogue played a significant role. Greek dialogue, often written in iambic trimeter, was delivered with a rhythmic cadence that enhanced audience comprehension of the text.[4]

Chanting and melodic recitation were also effective. Repetitive lyrical passages were chanted or sung to ensure the text was received over distances. The convention of the chorus delivered poetic dialogue in unison, using intonation to project commentary to the upper tiers of the theatre.[5]

Actors also practiced deliberate pauses and asides to allow dialogue to resonate with the audience. An aside is a dramatic technique involving deliberate pauses in delivery where actors address the audience directly to create moments of reflection. Pauses and asides after shocking scenes gave the audience a chance to absorb and process what just happened, while pauses after comedic lines gave the audience a chance to laugh.[5]

Resonating Masks

Masks had a symbolic role in rituals and theatre and were used from the Classical to the Hellenistic periods. They could identify a particular character, the character's social class, and occupation.[6] They were commonly used in the Dionysian Festivals and in the complex drama that came later.[6] Aside from the visual that the masks provided, they served an important role in actor intelligibility. Greek actors were trained to use masks with wide mouth openings that were cone-shaped. The cones functioned as rudimentary megaphones, directing sound toward the audience. Made of stiffened linen, cork, or wood, these masks enhanced midrange frequencies in the vocal range. The resonating mask amplified the volume of the voice, while the conical mouthpiece directed the sound of the voice. These masks worked synergistically with venue acoustics and actors' vocal techniques to enhance vocal clarity.[7]

Echea

Vitruvius describes bronze or clay "pots" (echea) that were tuned to absorb lower frequencies and thereby resonate the actors' voices. They were strategically placed in the theatre's seating area or near the stage, with openings facing the performance space. He suggests that they greatly increased the clarity of the actor's voice.[8]

Some modern studies opine that the effectiveness of echea was probably minimal at best.[9] It is important to consider that any additional aid to intelligibility, even a minimal one, contributed to the many acoustical techniques employed during this time period. Though their impact was likely minimal, echea reflected the Greeks' innovative approach to acoustics.

Acoustic Design

The acoustic properties of Greek and Roman theatres were critical to vocal clarity. Large theatres, built into hillsides, used semicircular seating and skene backdrops to project sound toward spectators. Vitruvius writes that the ancient Greek theatre venues were designed and built to have great acoustics, particularly three key elements: the "theatron" (seating area), "orchestra" (performance space), and "skene" (backdrop).[8]

The theatron has a semicircular seating area built into a hillside to leverage natural topography, with tiered rows divided into "kerkides" (wedge-shaped sections) by stairways for audience access.[1] Vitruvius notes that the theatron's curved layout enhances acoustics by focusing sound from the orchestra toward spectators, reducing distortion and amplifying the actors' voices naturally. He points out that the careful selection of the location of the building site was crucial. (For example, perhaps they avoided windy locations.)[8]

A skene (wooden backdrop wall) was added to the back of the orchestra area when dialogue became an important aspect of the program. It served to reflect actors' voices back to the theatron area to enhance audibility. "Skene" over time evolved into "Scene."[10]

According to Vitruvius, the Greek outdoor theatres were built in precise proportions to optimize the concepts of harmonic intervals, likely influenced by Pythagoras, to ensure the voice "strikes the ear with clearness and harmony." He discusses how the theatre's proportions should follow harmonic ratios to optimize sound reflection and avoid dissonance. In this respect, he likens the voice to a musical instrument, requiring a well-tuned environment.[8]

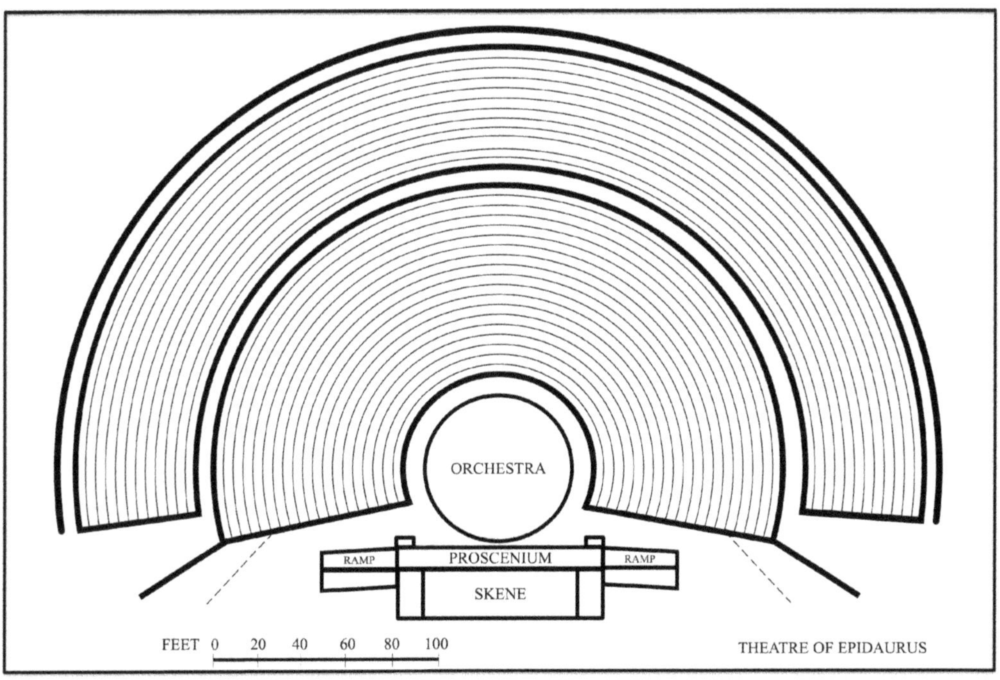

Figure 9.1 Block diagram of the Theatre of Epidaurus.

Drawing by the author.

Sound Effects

Ancient Greek theatre employed simple techniques to create sound effects to enhance their ritualistic stories of their gods and myths. They used sound effects as exposition and to indicate things like the entrance of a god or a demon.

The bronteion, described by Julius Pollux in his 2nd century AD *Onomasticon*, executed a thunder effect using air-inflated hide sacks filled with small stones struck against brass plates beneath the stage that created a large resonating boom. Later, the Romans reengineered their own thunder machine. Heron of Alexandria writes about a thunder machine which dropped weights through a metal or animal hide chute to produce a kettledrum-like sound. Romans also developed the "thunder-run," where large stones rolled down sloped wooden planks onto brass plates.[10]

Both Greek and Roman actors themselves contributed to soundscapes by vocally imitating animal noises and mechanical sounds like creaky hinges.[10] Actors considered it part of their craft and became quite skillful, creating all sorts of sounds from offstage utilizing their voice.

In addition to mechanical sound effects, Romans employed musical innovations like the hydraulus. This was a water-powered pipe organ. Theodoric the Great described its use in 5th century AD programs, producing "strange sounds" and "melodious" blasts with water-driven air pressure capable of projecting over vast distances. Archaeological finds confirm its advanced design with pipes, a keyboard, and perforated sliders. The hydraulus enabled Romans to craft dramatic musical effects, both elongated (legato) or brief (staccato).[10]

Contributions of Key Figures

The success of Greek and Roman acoustics was informed by scientific and philosophical advancements from a few key figures whose work revolutionized the culture and their theatre.

Aristotle (384–322 BCE)

Aristotle's acoustic theories provided a scientific foundation for understanding sound in theatre. He posited that sound is produced by the striking of objects, causing air to vibrate, and that it requires a medium like air or water to travel. He distinguished between actual sound and inert potential sound, laying the groundwork for later acoustic theories.[11] Aristotle's observations on echo reflection – where sound bounces off smooth surfaces like a mirror – explained how amphitheater designs can help achieve vocal clarity.[11] His *Poetics* also influenced theatrical structure, emphasizing clear dialogue to achieve mimesis (the imitation of representation: also known as acting) and catharsis, ensuring actors' voices conveyed meaning. Aristotle's ideas informed the intentional design of Greek theatres and of actors' vocal techniques, which greatly contributed to the goal of intelligibility.

Pythagoras (c. 570–495 BCE)

Pythagoras, a mathematician and philosopher, discovered that the pitch of a vibrating string depends on its length, with specific ratios producing harmonious intervals like the octave and perfect fifth. His Pythagorean tuning system likely influenced Greek architects' understanding of sound propagation, resulting in theatre designs with proportional layouts that optimized acoustics.[12] The semicircular seating and skene in theatres such as The Theatre of Epidaurus in southern Greece likely reflected Pythagorean principles of harmony, ensuring that voices were amplified naturally. Pythagoras' work supported the Greek goal of vocal intelligibility, creating meaningful sound experiences through scientific precision.

Vitruvius (c. 80–15 BCE)

Vitruvius, a Roman architect, detailed Greek theatre acoustics in *De Architectura*, emphasizing harmonic ratios, site selection, and resonating devices. He described how the theatron's curved layout and skene backdrop focused sound toward spectators, reducing distortion. Vitruvius also noted the use of *echea*, bronze or clay vessels tuned to resonate vocal frequencies, placed in niches to enhance clarity. His principles, applied in Roman theatres like the Theatre of Marcellus, ensured that actors' voices were intelligible in large venues. Vitruvius' work bridged Greek and Roman practices, reinforcing the importance of acoustic design in creating meaningful theatrical experiences.[8]

Chapter 9 Exercises

Exercise 1: Reflection on Acoustical Goals

Objective: Write a 200–300-word reflection on the main acoustical goals of the Greeks and Romans, how they pursued those goals, and how they can be applied to your own sound design projects. Focus on intentions, context, constraints, and choices.

Steps:

1. Identify the two main acoustical goals of Greek and Roman theater sound designs.

2. Describe specific techniques and technologies used to achieve these goals, considering the context of large open-air amphitheaters.
3. Explain how the intentions (e.g., audibility, emotional resonance) and constraints (e.g., venue size, ambient noise) shaped their choices.
4. Reflect on how these ancient practices can inform your own sound design projects. Choose a scene from a contemporary play. Explain how these ancient goals might be achieved in the scene.

Exercise 2: Sound Effects From Ancient Technologies

Objective: Write a 200–300-word reflection on how the Greeks and Romans used technologies like the bronteion thunder machine or hydraulus organ to add emotional impact in amphitheaters. How did one technology create meaning in a play's scene? Why was it effective? Suggest a modern sound design inspired by this technology. How would it affect today's audience?

Steps:

1. Select one ancient sound effect technology (e.g., bronteion thunder machine or hydraulus organ) and describe how it was used in Greek or Roman amphitheaters.
2. Analyze how this technology contributed to emotional meaning in a specific scene of a play, explaining why it was effective for the audience.
3. Propose a modern sound design inspired by this ancient technology, specifying the context (e.g., a contemporary play or performance).
4. Discuss how your proposed sound design would impact a modern audience, focusing on emotional and cultural resonance.

10

The Middle Ages

The Middle Ages began around the fall of the Western Roman Empire, roughly 476 AD, and lasted until the 14th or 15th century with the start of the Renaissance. After the fall of Rome, theatre was suppressed by some members of the Catholic Church due to its believed association with paganism and distractions from Christian piety. Yet, ironically, monks began experimenting with dramatic forms based on Bible stories in the 10th century. This evolved into a genre which was called *Quem quaeritis*, which translates to "whom do you seek?" It comes from a key question in the Easter liturgy, which dramatizes the moment when the women visit Jesus' tomb and find it empty. These liturgies were short, often sung, and performed in Latin by monks within church services, marking an early form of medieval drama. *Quem quaeritis* laid the groundwork for the dominant theatre throughout the rest of the Middle Ages.[1] By the 12th century, longer plays began to emerge, experimenting with mystery plays or miracle plays.

By the 12th century, liturgical plays at this time were scripted in the vernacular languages in France and England in order to reach a broader audience, although with some ecclesiastical oversight. By the 14th century, theatre expanded to many towns with morality plays, religious spectacles, and Corpus Christi pageants. Local residents performed, making these events part of community life. They reinforced religious ideas and promoted social behavior.[1]

While religious and morality plays dominated during this period, secular plays were also produced. Secular farces became significant in the late-medieval public squares and marketplaces. They often featured exaggerated humor. Farces like *Holy Deadlock, Shit for Brains*, and *The Farce of the Fart* contained

DOI: 10.4324/9781003519393-13

elements such as parodic saints or crude synonyms for body parts. These performances were part of a broader "Jongleuresque" tradition, where solo performers or small troupes entertained with poetry, songs, and comedic acts.[2]

Medieval theatre addressed the practicalities of actors being heard and understood in their own ways. They faced unique constraints due to the period's technological limitations, diverse performance venues, varied audiences' literacy levels, and the oversight of the Catholic Church. There were no significant advancements in acoustical theory comparable to the contributions of ancient Greek and Roman scholars. This period was marked by a focus on pragmatic, functional approaches to sound in the theatre and music.

Constraints

Lack of Permanent Theatres

Purpose-built, dedicated structures like the amphitheaters of ancient Greece and Rome did not exist. Instead, performances were staged in temporarily adapted or multipurpose spaces tailored to the type of play. Churches were used, and open-air venues such as town squares and wagon plays were used, which were affected by ambient noise, wind, urban noise, and even crowd noises.[3] Churches were often highly reverberant, which affected dialogue's clarity, so actors relied on over-enunciation and precise amounts of projection (not too much, not too little), which was sustained throughout the performance.

Variable Audience Sizes

The lack of permanent theatre structures resulted in audiences that ranged from 50 to 200 spectators in churches and from 500 to several thousand in town squares. This meant that actors had to employ varying techniques as they traveled from venue to venue. Smaller audiences required a more intimate delivery style, while larger crowds required actors to project over larger distances.[4]

Lack of Audience Literacy

Many of these spectators were unfamiliar with Latin, which was widely used in liturgical dramas, and ignorant of the complex theological concepts expressed in them. This required conventions such as rhythmic verse and repetition to address audience comprehension.[3]

Economic and Resource Constraints

Many early medieval troupes were small and nomadic bands of mimes or minstrels. These troupes earned meager income through donations and small fees, which restricted their ability to afford elaborate production. Most of their audiences had little disposable income, often prompting actors of these troupes to rely on barter in the form of food or lodging. This lack of funds hindered their production values such as costumes and scenery, as well as travel and even the basic needs of the actors. As a result, these troupes were often too small to perform conventions such as large choral chants that could help with delivering text to audiences.

Theatre Venues

Due to the lack of dedicated theatres, Medieval theatre venues varied widely. Most existing buildings temporarily accommodated performances, and so spaces often had to be adapted or required specific techniques to aid audibility.[5]

Churches and Cathedrals

Churches served as venues for religious theatre. Liturgical dramas were performed by members of the parish clergy, such as monks or choirboys. They used existing architecture near altars or choir areas, with small scenic structures (mansions or sedes) for specific locales. In small villages, audience sizes were small as well, while cathedrals like Notre-Dame in Paris could conceivably host hundreds. Larger congregations meant more money could be devoted to production values.[6]

High-vaulted ceilings and stone and marble walls and floors produced long reverberation. Research shows that churches often suffered reverberation decays of up to three seconds at midrange frequencies and up to five seconds at low frequencies.[7] This can make music sound more pleasing but can play havoc with speech clarity. Following storylines was challenging to plays that relied heavily on dialogue. This pushed performances toward visual and musical elements, shaping the evolution of medieval theatre.[7]

Town Squares and Market Places

These central locations facilitated large gatherings, making them ideal for community performances. Performing in open-air settings posed a different set of challenges but also advantages over churches. Surrounding buildings likely helped reflect sound, and the audiences could get very close to performers.[1]

There were two primary staging methods in these spaces. Fixed staging involved setting up mansions – small temporary scaffolds or platforms used for different locations in the play. Actors traveled from one platform to the other depending on the needs of the play. The other staging method involved pageant wagons.[1]

Pageant wagons traveled from town to town delivering episodic performances of biblical stories, such as the Crucifixion and Noah's Ark. They were drawn by horses or pushed by guild members. They were two-tiered carts with a lower compartment for storage or actor preparation and an upper platform for performance. Some had a canopy or third level where scenes set in heaven were acted out. Many employed trapdoors that enabled actors to rise or descend onto the stage, and pulleys lifted characters for ascensions.[1]

Great Halls and Courtyards

Morality plays were sometimes performed on portable raised platforms in noble or civic halls, or the innyard rooms within the courtyards of public inns, especially by professional or semiprofessional troupes. Some of the plays performed were written specifically for noble or courtly audiences. Nobles often allowed their household staff to attend. Audience sizes varied depending on the venue.[3]

For sound effects, thunder clappers or trumpets were staged on or near the wagons. To maintain audibility, actors used elevated positions and projected with clear, rhythmic verse. Canopies or wagon sides helped reflect sound. Actors routinely paused for crowd control, waiting for the noise to lull before continuing their lines.

Audibility Techniques

The acoustical characteristics of venues in the Middle Ages did not help actors much with ensuring the intelligibility of their lines. They employed many of the traditional techniques developed by Greek and Roman theatres.

Vocal Projection and Exaggerated Diction

Diverse acoustic environments required actors to adapt their projection to the venue they were performing in. Powerful projection was especially needed for mystery plays, often performed on pageant wagons, where actors needed to reach large crowds. Actors learned breath control and diaphragmatic techniques to project their voices in a controlled manner.[8]

Chanting and Singing

Performers would chant and sing to harmonize with the characteristics of reverberant spaces such as churches. This often enhanced the spiritual impact of the text.[2] Transitioning in and out of chanting, singing, and straight delivery of the text helped the audience to follow the story and making it easy for them to remember it.[2]

Repetition

Repeated phrases, actions, or motifs within plays were common devices used to reinforce key moments and themes.[2]

Pausing and Pacing

Actors occasionally addressed the audience directly, breaking the fourth wall to clarify themes or emphasize morals. In Everyman, the actor playing Death spoke directly to spectators to reinforce the play's lesson.[3]

Sound Effects

Fire and Pyrotechnics

Fire and pyrotechnics were used in the Middle Ages. Fire was used in the *hellmouth*. The hellmouth was a large stage prop representing the entrance to Hell, common in mystery cycle plays. Flames or sparks emanating from the mouth depicted damnation or demonic activity below.[3] Torches and bonfires were fueled by wood, pitch, or animal fat in many venues, including pageant wagons.

Pyrotechnics might include small flashes or sparks to represent miracles or angelic interventions. Gunpowder was sometimes ignited, creating explosions. Cannons were sometimes used for battles, thunder, and to represent earthquakes. These were rare due to cost and safety concerns but made a big impact with large audiences.[3]

Thunder

The sound of thunder was created by several different methods. One method was to cover wooden casks with skins or a cauldron covered with parchment skins. In his book *The Sounds of Theatre*, David Collinson describes a "Thunder Machine" using two large copper basins filled with rocks that were fashioned to a frame that allowed the basins to be rotated.[9] Varying versions of these machines were used when they could be afforded. Collinson also mentions the use of a thunder sheet during this time period.[9]

Whistling or Wind Machines

Simple mechanisms to evoke wind or storm effects were used in productions. Devices like large bellows and fabric sheets were shaken vigorously backstage to represent the sound of gusts. In some cases, stagehands might have swung ropes or rattled wooden boards to mimic the effects of wind.

Chapter 10 Exercises

Exercise 1: Recreating Medieval Vocal Projection Techniques

Objective: Understand and apply medieval actors' vocal projection and diction techniques to address acoustical challenges in varied performance spaces.

Instructions:

1. **Setup**: Choose two contrasting performance spaces on campus (e.g., a small classroom and a large outdoor courtyard or gym). These will simulate a church and a town square, respectively.
2. **Text Selection**: Select a short monologue (1–2 minutes) from a medieval play, such as *Everyman* or a mystery play.
3. **Performance**:
   - In the small indoor space, perform the monologue using controlled projection and exaggerated diction to combat reverberation (mimicking church acoustics). Focus on clarity without shouting, using breath control and diaphragmatic support.
   - In the outdoor space, perform the same monologue, projecting to reach an imagined audience of 200–500 people. Incorporate chanting or singing for key phrases to enhance audibility, as medieval actors did.
4. **Reflection**: Write a reflection comparing the challenges of each space. Address how you adjusted your vocal techniques, the effectiveness of chanting or repetition, and how the space's acoustics affected audience comprehension. Reference the chapter's discussion of church reverberation and open-air sound reflection.

Exercise 2: Designing Medieval Sound Effects

Objective: Create and test simple medieval-inspired sound effects to enhance a theatrical scene, addressing the period's technological and economic constraints.

Instructions:

1. **Scene Selection**: Choose a short scene (2–3 minutes) from a medieval mystery or morality play that includes environmental or dramatic effects (e.g., a storm in *Noah's Ark* or the hellmouth in a mystery

cycle). If unavailable, use a modern scene requiring similar effects (thunder, wind, or fire).

2. **Sound Effect Creation**:
   - Construct one sound effect device inspired by medieval techniques. Use only low-cost, readily available materials to reflect medieval economic constraints. Be as creative as you would like.
   - Test the device in a classroom or small performance space to ensure it produces the desired effect without electronic amplification.
3. **Performance**: Perform the scene in class, integrating your sound effect at appropriate moments. Ensure the effect is audible to an audience of 20–30 people, as in a small medieval venue. Coordinate with classmates to act out the scene while you operate the sound device.
4. **Reflection**: Write a reflection on the process. Discuss the challenges of creating the effect with limited resources, how well it enhanced the scene's atmosphere, and how it compares to modern sound design methods. Reference the chapter's examples of thunder machines or pyrotechnics.

11

The Renaissance – Transitioning to the Modern Theatre

The Renaissance, meaning "rebirth" was transformative for the arts and culture in Europe, from the 14th to 17th centuries. This period marked a shift from medieval religious dramas to secular plays with themes of the human experience influenced by the philosophy of humanism. Humanism, an intellectual movement of the time, focused on an individual's potential and scholarship. It was a response to the authority of the Catholic Church in the Middle Ages. It emphasized education, especially in the classical Greek and Roman texts. It sparked a cultural revolution that originated in Italy and spread across the continent of Europe and the British Isles. Shakespeare's plays aligned with humanism. The music of the era was influenced by humanism. Every theatrical element in this period – playwriting, theatre architecture, scenic, costumes, lighting, and sound – reflected the period's scholarly pursuits in the arts, sciences, and philosophy.

Theatrical Architecture and Acoustics

Humanism's revival of ancient Greek and Roman knowledge led architects to study classical theatre designs, which included specific acoustic considerations. Influenced by the rediscovery of Vitruvius's *De Architectura*, there was a trend to move from performances in predominantly temporary outdoor stages to permanent structures. One ancillary benefit meant an improvement of acoustical environments, although better acoustics was not the main goal.

DOI: 10.4324/9781003519393-14

Italy – Teatro Olimpico

Andrea Palladio's design for Teatro Olimpico, built in Vicenza in 1585, was informed by the desire to return to the lessons of the Greek and Roman theatre. Influenced by the writings of Vitruvius, the theatre was built with a proscenium stage and a classical semicircular auditorium that mimicked the open-air design of Greek amphitheaters. Directly in front of the stage was an orchestra area at the seating level, lower than the stage. The seating area surrounded the orchestra. This orchestra area refers to the classical architectural space modeled after ancient Roman theatre, rather than what we think of as a modern sunken pit for musicians. It served as an open performance or transition space, historically intended for the chorus, for dignitaries, and for integrated audience viewing, without a dedicated pit as seen in later opera houses.[1]

This design, drawn from Vitruvius' principles emphasizing sightlines and immersion, uses the orchestra's radius to influence the cavea's proportions. The cavea is approximately 1.5 times the orchestra's diameter. The building of Teatro Olimpico serves as a significant testimony to the dedication and commitment to cultural advancement by returning to the traditions of the classical Greek and Roman theatre.

England

Playhouses in England were built with cost in mind. Audience capacity and low building expenses were prioritized over specialized design. Features like timber and thatched roofs were chosen for their economy, not for acoustical qualities. Acoustical functionalities did exist, even if purely incidental.

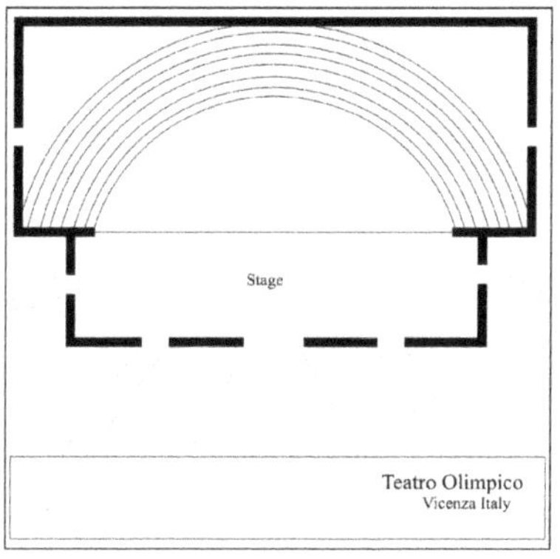

Figure 11.1 Block diagram of Teatro Olimpico, Vicenza, Italy.

Drawing by the author.

English Theatre Architecture and Layout

Amphitheaters

Amphitheaters such as The Theatre, The Globe, and The Red Lion typically were polygonal in shape, had brick foundations, and were built with timber frames with plaster infill. The roof was either thatched, like The Globe, or tiled. They featured an open yard where the cheapest patrons stood, surrounded by tiered galleries where the main audience sat. The elevated stage was usually large and protruding, with a tiring-house built upstage center.[2]

The tiring house, positioned at the back of the stage, was a three-tier setup. The ground level had a raised inner stage, more suited for smaller scenes. The next level was a central playing area, and the top tier featured a central gallery for musicians who played behind curtains. On this level were curtained off spaces to the right and left used as dressing areas and storage.[2]

The first playhouse built to last was The Theatre. The Theatre was built by James Burbage in London in 1576 and accommodated up to 1,500 audience members.[3] It was built specifically for the Leicester's Company, also known as the Leicester's Men. In 1594, William Shakespeare's company, known at the time as Lord Chamberlain's Men, began using The Theatre as their main space, although not exclusively. They also shared performances at the nearby Curtain playhouse, especially as tensions with the landlord escalated toward the end of the land lease. Lease disputes led to the dismantling of the space in 1598 as the lease expired.

In 1599, The Globe Theatre was constructed in London from The Theatre's timbers. It featured areas for the "Heavens" and "Hell." The Heavens was located at the roof level above the stage, and the area beneath the stage was Hell, accessible through a trap door on the stage.[2]

There were unplanned acoustical advantages to this design. The polygonal layout created a natural acoustic focus, with sound radiating to the surrounding audience. This minimized sound dispersion as compared to a rectangular house, where sound could be lost in corners, and low acoustical standing waves and nodes can develop. A standing wave is a sound wave that remains fixed in place within a confined space, like a room, due to repeated reflections off the boundaries. A node is a point along a standing sound wave where the displacement of the medium remains zero, meaning that there is no vibration or net movement at that location.

The Globe also had a canopy over the stage, which likely acted like a horn, reflecting sound that would otherwise be lost to the open sky. The stage likely acted as the bottom side of that horn. The tiered galleries slightly mimicked the slope of the Greek amphitheaters, which allowed more audience members to be closer to the stage than that of an audience area that had no slope or tiers.

There were several acoustical constraints for intelligibility in this type of layout.

- Weather and noise: Wind, rain, and ambient urban noise could literally drown out the actors' voices at times. The Globe, cleverly, was built just outside London's boundaries.

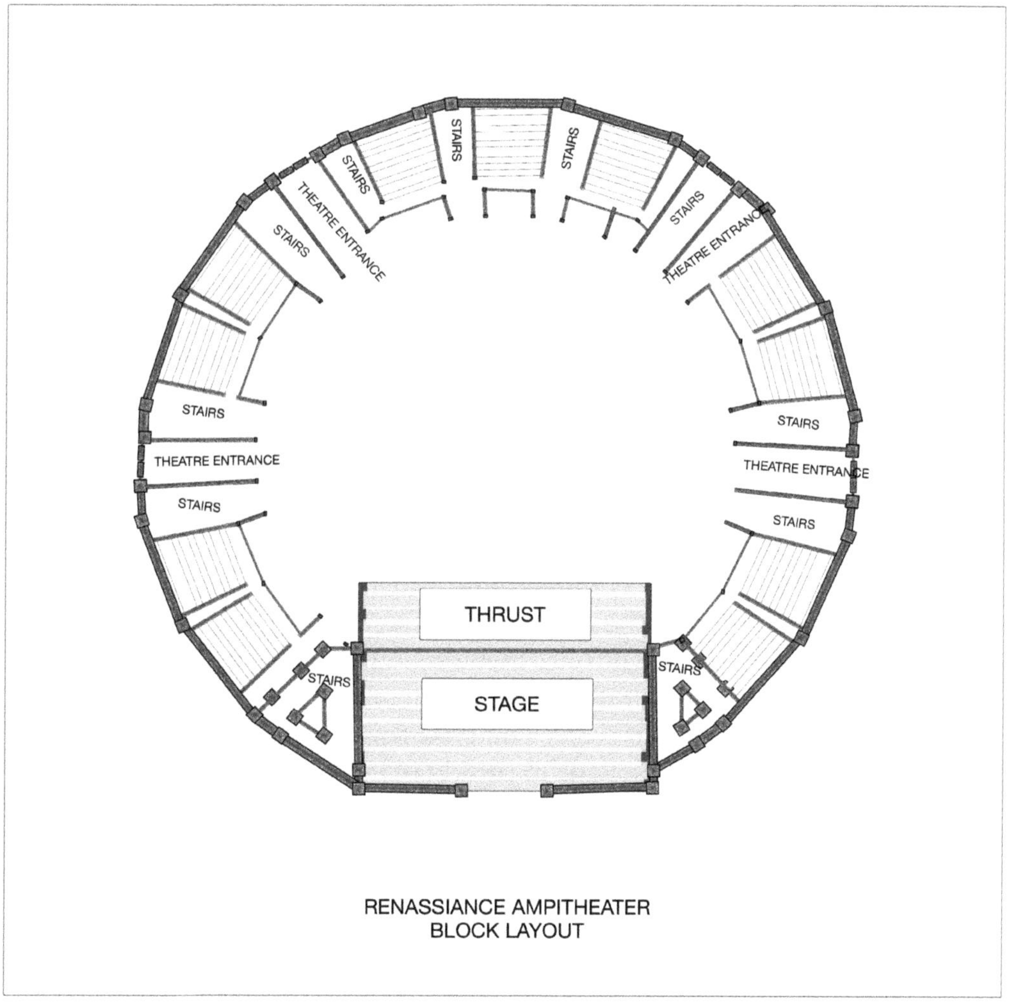

Figure 11.2 Block diagram of a typical Renaissance amphitheater.

Drawing by the author.

- Large audiences: The Globe theatre was estimated to be able to accommodate over 3,000 audience members. This required actors to project loudly to reach the back of the house.
- Shape of the galleries: The rounded shape of the galleries meant that there was a wide dispersal of audience members. Actors had to keep to precise blocking patterns to ensure everyone followed the story. Turning to the right, for example, at the wrong time meant facing away from a significant portion of the audience.
- Pillars supporting the "Heavens" obstructed the view and sound for a few seats. They could at times also create unpredictable patterns of sound propagation.
- Crowd noise: Groundlings in the yard were mobile and moved around, potentially creating a disturbance. Chatter from that area also interfered with dialogue.

Hall-Shaped Playhouses

Hall playhouses were rectangular and often converted from existing buildings such as barns. Stages were narrower and set across the shorter end of the hall. The tiring-house for this type of playhouse typically had multiple doors, a music room above, and sometimes a balcony. Trapdoors and *Heavens* for descents were present. Wealthier patrons sat closer to the stage, while the galleries were cheaper. Sometimes boxes were built house left and house right. Programs were sometimes performed in Court theatres, where temporary stages were erected. They shared the same shape as hall-shaped playhouses, and so they are placed in this category.

There were a few advantages of sound propagation as compared to amphitheaters. Since hall-shaped playhouses were indoor venues, the actors' vocals were not required to compete with much ambient noise, and sound could reflect off the walls, the floor, and the ceiling. The stages were typically narrower, as in The Blackfriars theatre, and concentrated sound output onto the audience area. Generally, audience members were closer to the stage, especially in the pricier front seats. The placement of the music room above the stage positioned the musicians to project over the audience, ameliorating the tendency to overpower the actors' voices.[2]

The acoustical constraints included:

- Crowded stage: The number of cast members on the smaller stage could exceed 15 or more, turning these actors into obstructions. The narrower stage hindered actors' movements and stage blocking, forcing them to deliver their lines from confined positions.

- Audience noise: While proximity to the stage was a benefit, it was often a disadvantage, as wealthier (and therefore socially prominent) patrons close to the stage were often vocal and could be quite distracting at times.
- Instrumentation: Even though the musicians were placed in the upper room above the stage, they were forced to use softer instruments and softer playing styles so as not to compete with the dialogue if not carefully balanced.
- The square shape of the hall had a high incidence of reflection, creating standing waves and nodes.

Sound Effects and Music in Renaissance Theatre

To create sound effects, actors and craftsmen employed manual techniques, pyrotechnics and ordinance, vocal mimicry, and elaborate machines. Dedicated venues allowed them to image sound effects a bit more strategically for a better effect. Conventions evolved with certain sound effects relating to certain actions or ideas, such as a trumpet fanfare that introduced important characters' entrances or swords clanking offstage to indicate combat.

Environmental Sound Effects

We see thunder runs used quite effectively in the Renaissance. Cannonballs, 4–6 inches in diameter, were rolled down wooden troughs, producing a rumbling thunder. In the Globe, stagehands operated these in the "Heavens," which was located directly above the thrust stage.[2] Thunder sheets were also used extensively for this purpose.[3]

For wind, bellows were used that were 2–3 feet in length. Another device used fabric strips to create the sound of whooshing wind, used in *The Tempest*. In the Blackfriars, stagehands operated these in the tiring house. The Globe's backstage openings project wind sounds as well.[2] Another wind machine contained wooden drums with slatted surfaces to produce howling winds.[2]

Water: Metal or wooden containers about 1 or 2 feet in diameter sloshed water to mimic waves, as in *The Tempest*. Teatro Farnese's trapdoors and tiring house placement created directional aquatic sounds reflecting off the wooden construction.[3]

Lightning: Small paper tubes with gunpowder, called squibs, created pops for lightning. In the Globe's heavens, fire masters ignited squibs for overhead imaging.[2]

Battle and Action Sound Effects

Trumpets and drums signaled battles, as in *Macbeth*'s alarums. In open-air courtyard theatres like The Globe, musicians played in the upper galleries or above the stage, ensuring the sound carried across the open yard to the spectators. They sometimes performed in the tiring-house for offstage sound cues through backstage openings.[2]

Sword Combat: Actors used metal swords or props to make percussive fight sounds onstage, and offstage, a stagehand or two might use them from the tiring house to suggest distant combat.[4]

Chambers or small cannons created booms, as in *Henry V*. The Globe's courtyard placement projected sound all over the theatre building.[3] Infamously, a cannon misfire resulted in the Globe burning to the ground.[2]

Supernatural and Symbolic Sound Effects

Bells and Chimes: Bells produced supernatural tones. In the Globe's understage "Hell," bells carried through the open structure.

Whistles: Small whistles and birdcalls were often used to create ghostly sounds.[2] Blackfriars' tiring house had built-in resonating pipes through which the sounds of the whistles would travel.[2]

Shakespeare's Conventions

To get a sense of the conventions used during the Renaissance in the English theatre, we can look at Shakespeare's writing. This is a list of all sound effects in selected Shakespeare that represent the bulk of his work. These sound effects come from: *A Midsummer Night's Dream, Cymbeline, Hamlet, Henry IV, Part 1, Henry V, Henry VIII, Julius Caesar, King Lear, Macbeth, Much Ado About Nothing, Othello, Richard III, Romeo and Juliet, The Comedy of Errors, The Merchant of Venice, The Merry Wives of Windsor, The Tempest, The Two Gentlemen of Verona, The Winter's Tale, Twelfth Night,* and the narrative poem *Venus and Adonis*.[5]

Alarum

A call to arms signaling battle or urgent action, the *alarum* created chaotic tension in military scenes. It was produced with rapid drum rolls or trumpet blasts by musicians offstage or in the tiring-house, often with shouts or weapon clashes for added chaos. This sound appears in *Henry V, Julius Caesar, Macbeth, King Lear,* and *Henry IV, Part 1*. Alarums are common in histories and tragedies, where battles and political upheavals drive the narrative.

Alarum and Retreat Within

Used to depict offstage battle sounds followed by a withdrawal signal, this sound heightened the drama of combat sequences. It was created with an initial *alarum* of trumpets and drums, followed by a slower trumpet fanfare or distinct drum rhythm for retreat, performed backstage. It appears in *King Lear*. Such sounds are typical in many Shakespearean tragedies with climactic battles.

Alarums (Combat)

Conveying chaotic sounds during staged combat, these *alarums* intensified fight scenes. They were produced with rapid, irregular drumbeats and possible trumpet calls backstage, with actors clashing weapons. This sound is found in *King Lear*. Combat alarums are common in many Shakespearean tragedies and histories featuring duels or warfare.

Ambient Crowd Noise

Adding atmosphere to lively gatherings, *ambient crowd noise* mimicked the murmur of a crowd. It was created by actors or stagehands offstage, vocalizing softly or rustling props like cloth. This effect appears in *Much Ado About Nothing*. Such sounds are common in comedies, where festive or communal scenes abound.

Ariel's Sleep Music

An enchanting melody to lull mariners to sleep, this music created a magical atmosphere. It was performed with soft lutes, viols, or a sung lullaby by Ariel, played from the gallery or onstage. This sound is exclusive to *The Tempest*. Such ethereal music is typical in romances, emphasizing supernatural elements.

Ariel's Song

A joyful song about freedom sung while attiring Prospero, this music underscored liberation. It was sung by a boy actor, possibly with lute accompaniment, performed onstage or from the gallery. This sound appears in *The Tempest*. Songs of this nature are common in romances and comedies, enhancing emotional or celebratory moments.

Ariel's Storm Effects

Magical fire and thunder sounds amplified the supernatural storm's impact. They were created using thunder sheets or rolling cannonballs for thunder, with torches or pyrotechnics for fire effects, performed backstage. This effect is unique to *The Tempest*. Such effects are prevalent in romances and tragedies with supernatural themes.

Ariel's Tabor Music

Rhythmic beats to enchant conspirators, this music added a mystical quality. It was produced with a tabor drum, possibly with a pipe for melody, played onstage or offstage. This sound is found in *The Tempest*. Rhythmic music like this is common in romances, supporting magical or ritualistic scenes.

Bell

Signifying alarms, time changes, or ceremonies, bells added dramatic weight. They were created by ringing a handbell or metal gong backstage by a stage-hand, with varying sizes for tones. This sound appears in *Macbeth* and *Henry V*. Bells are common in tragedies and histories, marking significant events or transitions.

Caliban's Voice Within

An offstage voice responding to Prospero, this sound created mystery. It was produced by an actor speaking from backstage or through a trapdoor. This effect is exclusive to *The Tempest*. Offstage voices are typical in romances, enhancing supernatural or hidden actions.

Cannon Shot

Indicating large-scale battles or significant offstage events, *cannon shots* heightened military drama. They were simulated by rolling a cannonball across a wooden floor (thunder run), striking a large drum, or using pyro-technic squibs. This sound appears in *Henry V* and *Henry VIII*. Cannon shots are common in history, emphasizing grand battles.

Chambers Discharged

Representing gunfire or explosive sounds, this effect punctuated dramatic moments. It was created by igniting gunpowder in a controlled device or using a wooden clapper for a loud bang. This sound is found in *Henry VIII*. Such effects are typical in histories and tragedies with violent or ceremonial scenes.

Clashes of Swords

Representing combat or duels, *sword clashes* intensified fight scenes. They were produced by actors striking blunted stage weapons or stagehands hit-ting metal objects like swords or iron bars backstage. This sound appears in

Romeo and Juliet, Hamlet, Richard III, Macbeth, and *King Lear*. Sword clashes are common in tragedies and histories with dueling or warfare.

Clock Striking

Mimicking a clock to indicate time passage, often at night, this sound set a temporal context. It was created by striking a bell or gong with a mallet or ringing a handbell multiple times backstage. This effect is found in *Macbeth*. Clock sounds are typical in tragedies, enhancing suspenseful or nocturnal scenes.

Clubs, Bills, and Partisans

Depicting citizens shouting during a street brawl with weapon sounds, this effect created chaos. It was produced by actors striking wooden or metal props (clubs, staffs, or blunted swords) against each other or shields, choreographed for rhythmic clanking. This sound appears in *Romeo and Juliet*. Such effects are common in tragedies with civil unrest.

Cock Crows

Signaling dawn or supernatural events, the *cock crow* added an eerie tone. It was mimicked by an actor or stagehand vocalizing a rooster's crow or using a whistle/pipe. This sound is found in *Hamlet*. Cock crows are typical in tragedies with supernatural or transitional moments.

Confused Noise of Voices

Offstage cries of panic during a shipwreck, this sound heightened chaos. It was created by actors shouting backstage in unison or staggered. This effect is exclusive to *The Tempest*. Such sounds are common in romances, emphasizing dramatic upheavals.

Cornets

Ceremonial music for royal entrances or court scenes, *cornets* provided a regal tone. They were played live by musicians using high-pitched brass cornets in the music gallery or offstage. This sound appears in *Henry VIII* and *Cymbeline*. Cornets are common in histories and romances with courtly settings.

Cry of Help

Shouts like "Help!" or "Murder!" conveyed shock or summoned assistance, amplifying tension. They were vocalized by actors onstage or offstage, amplified by theater acoustics. This sound is found in *Macbeth* and *Othello*. Such cries are common in tragedies, highlighting distress or violence.

Cry of Murder

Shouts of "Murder!" simulated panic in violent scenes, creating chaos. They were vocalized by actors onstage and offstage. This sound appears in *Macbeth*. *Murder cries* are typical in tragedies with intense, violent plots.

Cry Within

Offstage shouts or wails signaling distress or approach, this sound created suspense. It was produced by actors or stagehands offstage, producing loud vocal clamor, amplified by theater acoustics. This sound appears in *Macbeth* and *Romeo and Juliet*. Offstage cries are common in tragedies and romances with dramatic offstage action.

Dance of Satyrs

A rustic dance during a festival, accompanied by music, this performance added a lively, pastoral tone. It was created with musicians playing drums, pipes, or shawms for a pastoral tune, with actors' stomping or props like bells/clappers adding percussion. This sound appears in *The Winter's Tale*. Such dances are common in romances and comedies, enhancing festive scenes.

Dead March

Somber procession music for a tragic finale, this sound underscored mourning. It was created with a slow, muffled drumbeat, possibly with muted trumpets or recorders, playing a dirge-like rhythm. This sound is found in *King Lear*. *Dead marches* are typical in tragedies, marking solemn conclusions.

Drum

A military signal for army movements or battles, drums provided a martial rhythm. They were played live by drummers using snare, tabor, or bass drums with wooden sticks. This sound appears in *Henry V*, *Julius Caesar*, *Henry IV, Part 1*, and *King Lear*. Drums are common in histories and tragedies with military themes.

Drum Afar Off

Distant military drumming suggested an approaching army, building tension. It was created by softly striking large military drums backstage, possibly muffled. This sound is found in *King Lear*. Such sounds are typical in tragedies and histories with offstage military action.

Drum and Colors

Military drums for army entrances, this sound provided a rhythmic marching beat. They were played by drummers accompanying soldiers, producing a steady cadence. This sound appears in *King Lear*. Such effects are common in histories and tragedies with formal military scenes.

Flourish

A ceremonial fanfare for royal entrances or exits, *flourishes* symbolized grandeur. They were created with a short triumphant melody by trumpeters using natural trumpets or cornets in the gallery or tiring-house. This sound appears in *Henry V*, *Henry VIII*, *King Lear*, *Cymbeline*, and *Henry IV, Part 1*. Flourishes are common in histories and tragedies with royal or noble characters.

Hautboys

Reedy music for courtly or eerie scenes, *hautboys* added a distinctive tone. They were played by musicians using hautboys (early oboes) or shawms in the music gallery or onstage. This sound appears in *Macbeth* and *Cymbeline*. Hautboys are common in tragedies and romances with supernatural or courtly atmospheres.

Horn (Hunting Horn)

Hunting calls signaling nobility, transitions, or supernatural atmosphere, horns created a vibrant, outdoor tone. They were played offstage using hunting horns or trumpets, producing loud, resonant blasts. This sound appears in *Venus and Adonis* and *The Merry Wives of Windsor*. *Hunting horns* are common in comedies and romances with pastoral or supernatural elements.

Hounds' Cry/Baying

Sounds of hunting dogs during a chase, this effect evoked a lively pursuit. It was created by a performer vocalizing sharp barks or howls, with offstage actors yelping or a horn simulating hunting calls. This sound appears in *Venus and Adonis*. Such sounds are typical in romances and comedies with hunting scenes.

Howling (Hound)

A mournful hound sound conveying sorrow, this effect added emotional depth. It was produced by a performer creating a low, mournful howl, with an offstage actor vocalizing a prolonged, eerie howl. This sound appears in *Venus and Adonis*. *Mournful howls* are common in romances, enhancing pathos.

Hunters and Dogs Noise

Barking dogs and hunting horns during a chase, this sound created a chaotic hunting atmosphere. It was produced by actors mimicking barks and trumpets or horns for hunt sounds, played offstage. This sound appears in *The Tempest*. Such effects are common in romances with dynamic outdoor scenes.

Knock

Indicating someone seeking entry, building suspense, or announcing arrival, knocks heightened anticipation. They were created by a stagehand or actor striking a wooden door, panel, or hollow box with a fist or mallet, performed offstage or on a visible door prop. This sound appears in *Macbeth*, *Much Ado About Nothing*, and *The Merry Wives of Windsor*. Knocks are common in tragedies and comedies, signaling dramatic entrances or suspense.

Lions Roar

A comical or dramatic animal sound in a play-within-a-play, this *roar* added humor or tension. It was vocally produced by an actor with a loud, guttural sound, amplified by theater acoustics or hand-cupping. This sound appears in *A Midsummer Night's Dream*. Such sounds are typical in comedies, enhancing playful or theatrical moments.

March

Processional music for military or ceremonial scenes, marches provided a rhythmic structure. They were created by a small ensemble of drums and wind instruments (e.g., fifes, trumpets, shawms) in a rhythmic style. This sound appears in *Henry V*, *Julius Caesar*, and *Henry IV, Part 1*. Marches are common in history, underscoring military movements.

Masque Music

A masque is a form of courtly entertainment popular in that era. This music created a ceremonial atmosphere. It was performed by a consort of lutes, viols, and recorders, with boy actors singing onstage or from the gallery. This sound appears in *The Tempest*. *Masque music* was typical in romances, enhancing elaborate spectacles.

Master's Whistle

A nautical whistle signaling commands to mariners, this sound set a maritime context. It was created with a Boatswain's pipe or similar whistle blown by an actor. This sound appears in *The Tempest*. Such whistles are common in romances with seafaring themes.

Music

Melodic interludes for mood, supernatural, or celebratory scenes, including wedding/funeral music or statue awakening, this music enriched emotional depth. It was performed by a small ensemble using lutes, viols, recorders, or shawms, playing soft melodies from the gallery or offstage, with styles like pavanes or Elizabethan melodies. This sound appears in *Romeo and Juliet*, *The*

Winter's Tale, *Twelfth Night*, *Much Ado About Nothing*, *The Two Gentlemen of Verona*, *King Lear*, *The Tempest*, and *Cymbeline*. Music is common in comedies and romances, enhancing festive or emotional scenes.

Neighs (Horses)

Equine sounds for a horse, this effect added realism to outdoor scenes. It was created by a performer vocally mimicking high-pitched whinnies, with a stagehand using vocal imitation or a whistle offstage. This sound appears in *Venus and Adonis*. *Horse sounds* are typical in romances with pastoral settings.

Noise of Hunting

Chaotic hunting sounds accompanying a dispersal or chase, this effect created a lively atmosphere. It was produced with combined horn blasts, drum rolls, wooden clappers, and offstage shouts to simulate a hunt's clamor, performed within for distance effect. This sound appears in *The Merry Wives of Windsor*. *Hunting noises* are common in comedies and romances with outdoor or supernatural scenes.

Noise Within

Crowd sounds or commotion, such as shouting for an entrance, this effect suggested offstage activity. It was created by offstage actors or stagehands shouting and stomping or banging wooden surfaces. This sound appears in *Romeo and Juliet*. Such noises are common in tragedies and comedies with lively crowd scenes.

Non Nobis and Te Deum

Sacred choral hymns for post-battle gratitude, these songs conveyed reverence. They were sung by a small choir or actors, possibly with lute or organ accompaniment, often a cappella. This sound appears in *Henry V*. *Sacred music* is typical in histories, marking solemn victories.

Outcry

Communal vocal uproar shouting names, this sound mimicked a chaotic crowd. It was created by actors or stagehands offstage shouting names in a chaotic, overlapping vocal effect. This sound appears in *Romeo and Juliet*. Outcries are common in tragedies with intense public scenes.

Retreat

A military signal for withdrawal, this sound marked a strategic shift. It was created by trumpeters playing a measured fanfare on natural trumpets in the gallery or tiring-house. This sound appears in *Henry IV, Part 1*. *Retreat signals* are common in histories and tragedies with battle scenes.

Screech-Owl Screeching

An ominous sound for death or supernatural atmosphere, this *screech* heightened tension. It was mimicked vocally by an offstage actor/stagehand with a shrill cry or using a whistle. This sound appears in *Macbeth*. *Screech-owl sounds* are typical in tragedies with dark, supernatural elements.

Sennet

A formal trumpet fanfare for royal processions, *sennets* conveyed grandeur. They were played by trumpeters using natural trumpets, producing a stately melody from the gallery or offstage. This sound appears in *King Lear*. Sennets are common in histories and tragedies with royal entrances.

Shot (Ordnance)

Simulating a peal of gunfire for battles or salutes, this sound amplified military action. It was created by rolling a cannonball or striking a drum multiple times to mimic gunfire. This sound appears in *Henry V*. *Ordnance shots* are common in histories with large-scale battles.

Shout

Crowd or group exclamations for enthusiasm or joy, shouts added a communal energy. They were created by actors or stagehands vocalizing loudly offstage or from the audience area to simulate a crowd. This sound appears in *The Two Gentlemen of Verona*. Shouts are common in comedies and histories with celebratory or public scenes.

Shouting and Sounds of Fighting

Offstage noises representing pursuit or combat, this sound created a sense of action. It was produced by actors shouting chaotically backstage, striking wooden sticks or swords together, with possible small drum rhythms. This sound appears in *The Two Gentlemen of Verona*. Such sounds are common in comedies and tragedies with offstage conflict.

Singing (Fairies)

A scornful rhyme sung by fairies to mock behavior, this song added an eerie, moralizing tone. It was sung live by actors (boys as fairies) in unison or harmony, likely a cappella or with lute or recorder support. This sound appears in *The Merry Wives of Windsor*. *Fairy songs* are typical in comedies and romances with supernatural elements.

Singing (Mistress Quickly)

A brief snatch of a song while alone, this performance added character depth. It was sung live by the actor (likely a boy), possibly a cappella or with a lute, using a familiar ballad refrain, projected clearly for audibility. This sound appears in *The Merry Wives of Windsor*. Such solo songs are common in comedies, enhancing character moments.

Soldiers Music

A solemn martial tune for military scenes, this music underscored discipline. It was performed by a consort of trumpets, drums, and possibly fifes, from the gallery or onstage. This sound appears in *Henry V*. *Soldiers' music* is common in histories, reinforcing military themes.

Solemn and Strange Music

Eerie music for supernatural or dramatic scenes, this sound created an otherworldly atmosphere. It was performed by viols or cornets playing dissonant, minor-key music from the gallery or onstage. This sound appears in *The Tempest*. Such music is typical in romances and tragedies with supernatural tones.

Solemn Music

Mournful or restorative music for funerals or resolutions, this sound evoked solemnity or healing. It was performed by lutes, viols, recorders, or an organ playing slow, harmonious melodies in the gallery or onstage. This sound appears in *Cymbeline* and *The Tempest*. *Solemn music* is common in romances and tragedies, marking emotional climaxes.

Song

Musical interludes for emotional depth or plot, including ballads, satirical, or mournful songs, these performances enriched the narrative. They were sung by actors a cappella or with lute/viol accompaniment, using Elizabethan ballad tunes, with styles varying by play. This sound appears in *The Winter's Tale*, *Twelfth Night*, *Venus and Adonis*, *Much Ado About Nothing*, and *The Tempest*. Songs are common in comedies and romances, adding emotional or thematic resonance.

Song and Dance

A ritualistic performance for magical or celebratory resolutions, this combined singing and dancing for a festive effect. It was created by actors singing in harmony with lutes, viols, or hautboys, paired with choreographed dance (e.g., pavane). This sound appears in *A Midsummer Night's Dream*. Such performances are common in comedies and romances, enhancing magical or joyful resolutions.

Strange, Hollow Noise

A sound accompanying vanishing spirits, this effect created a mystical exit. It was produced with irregular drums or gongs and discordant shawms, played backstage. This sound appears in *The Tempest*. *Strange noises* are typical in romances, amplifying supernatural transitions.

Strange Noises

Roaring, shrieking, howling, and jingling chains, these sounds created a chaotic, eerie atmosphere. They were produced with thunder sheets, whistles, vocal howls, and rattled chains backstage. This sound appears in *The Tempest*. Such noises are common in romances and tragedies with supernatural elements.

Tempestuous Noise

Thunder and lightning sounds during a storm, this effect heightened the dramatic upheaval. It was created with thunder sheets or rolling cannonballs for thunder and pyrotechnics or reflective surfaces for lightning. This sound appears in *The Tempest*. *Tempestuous noises* are typical in romances and tragedies with stormy or supernatural settings.

Thunder

Representing storms, divine intervention, or horse hooves, thunder added a foreboding tone. It was created by rolling cannonballs down wooden troughs, shaking thunder sheets, or beating drums with soft mallets; for hoof sounds, a drum was struck, or a heavy object rolled. This sound appears in *Macbeth* and *Venus and Adonis*. Thunder is common in tragedies and romances with dramatic or supernatural scenes.

Thunder and Lightning

Creating a supernatural atmosphere for ominous scenes, this effect intensified tension. It was produced with thunder via cannonballs or thunder sheets and lightning via burning rosin powder or squibs for flashes. This sound appears in *Macbeth*. Such effects are typical in tragedies with dark, mystical themes.

Trumpet Sound

Heralding arrivals, battles, or challenges, *trumpet sounds* added a bold, ceremonial tone. They were played live by musicians using natural trumpets, producing a heraldic sound from the gallery or offstage. This sound appears in *Henry V*, *Henry VIII*, *King Lear*, *Cymbeline*, and *Henry IV, Part 1*. Trumpets are common in histories and tragedies with royal or military scenes.

Trumpets (Combat)

Fanfares to start combat, these trumpets heightened the drama of fight scenes. They were played by trumpeters in the gallery using sharp, ceremonial notes. This sound appears in *King Lear*. *Combat trumpets* are typical in tragedies and histories with staged battles.

First Trumpet

The first of three trumpet calls for a challenge; this sound initiated a ceremonial duel. It was a short blast by a trumpeter in the gallery, signaling the challenge's start. This sound appears in *King Lear*. Such *trumpet calls* are common in tragedies with formal combat scenes.

Second Trumpet

The second trumpet call for a challenge, this sound continued the ceremonial duel. It was another short trumpet blast, maintaining the ritualistic tone. This sound appears in *King Lear*. *Sequential trumpet calls* are typical in tragedies with structured challenges.

Third Trumpet

The third trumpet call, with a response, this sound marked a combatant's arrival in a challenge. It was a third blast, followed by a response from another trumpeter. This sound appears in *King Lear*. Such calls are common in tragedies, finalizing ceremonial combats.

Chapter 11 Exercises

Exercise 1: Designing Sound for a Shakespearean Scene

Objective: Understand how sound effects and music in Renaissance theatre create meaning, enhance narrative, and apply this understanding to design a soundscape for a specific scene.

Instructions:

1. **Select a Scene**: Choose a scene from one of Shakespeare's plays. Focus on a scene with prominent sound cues, such as a battle, a supernatural event, or a ceremonial entrance.
2. **Analyze the Scene**: Identify the key actions, emotions, and themes in the scene. Consider how sound can enhance these elements.
3. **Research Historical Techniques**: Refer to the chapter's descriptions of Renaissance sound and music. List at least three historically

appropriate sound effects or musical elements you could use, citing specific examples from the chapter.

4. **Design a Soundscape**: Create a sound plot for the scene, detailing:
   - **Sound Cues**: For each cue, describe the sound, its source (e.g., onstage, offstage, tiring-house), and the technique used to produce it.
   - **Meaning and Effect**: Explain how each sound cue contributes to the scene's meaning.
   - **Acoustical Considerations**: Given the Globe theatre's architecture, describe how the sound will project to the audience, along with any constraints.
5. **Reflection**: Write a reflection on how your sound design choices create meaning in the scene. Discuss how the sounds reflect Renaissance Humanism and how they align with the period's theatrical conventions.

Exercise 2: Acoustics and Sound Placement in Renaissance Theatres

Objective: Explore how the Globe theatre's architecture and acoustics influence sound design decisions and create meaning in Renaissance performances.

Instructions:

1. **Select a Sound Effect or Music Cue**: Choose one sound effect or musical element from the chapter's list of Shakespearean conventions. Describe how the sound would be produced.
2. **Analyze Acoustical Impact**:
   - **Sound Propagation**: Explain how the theatre's architecture affects the sound's projection.
   - **Placement**: Decide where the sound should originate to maximize its impact and meaning. Justify your choice based on the theatre's acoustics and the sound's narrative role.
   - **Challenges**: Identify at least two acoustical constraints and propose solutions.
3. **Create a Diagram**: Draw a simple diagram of the theatre's layout, labeling key areas (e.g., stage, tiring-house, galleries, Heavens). Mark where your sound originates and how it travels to the audience.
4. **Write a Case Study**: Describe your sound effect or music cue, its placement, and how the theatre's acoustics shape its meaning. Connect your analysis to the chapter's discussion of Renaissance Humanism and classical influences on theatre design.

12

The Italian Renaissance – The Birth of Opera

A primary goal behind the creation of opera was to ensure performers could be clearly understood. This new genre combined vocal and instrumental music, drama, and sometimes dance to create a new form of storytelling. The enduring goal of ensuring that the actors' lines and lyrics were clearly audible shaped the composition and orchestration techniques such as the development of monody and recitative and the strategic use of orchestration and sound effects to enhance emotional impact.

The Invention of Opera

In the late 15th century, an informal circle of musicians, poets, and intellectuals gathered in Florence, Italy. They called themselves the Florentine Camerata (camerata is Italian for "room" or "chamber," describing a group of people who gather in a room for a common purpose). They gathered under the patronage of the prominent Giovanni de' Bardi, meeting in a room in his palace. Inspired by humanism, they discussed an intention to revive the dramatic power of ancient Greek tragedy as described by Aristotle in his *Poetics*, which they believed was entirely sung. Captivated by Aristotle's concept of catharsis – a purification of emotions through empathy with dramatic characters – they hypothesized that Greek tragedy's impact was derived from song, not spoken words. This led to the idea of a new form of theatre that could combine continuous music with drama to maximize the cathartic effect.[1]

DOI: 10.4324/9781003519393-15

Among the Florentine Camerata participants was Vincenzo Galilei, who coincidentally was the father of astronomer Galileo Galilei. Vincenzo Galilei was a musician and composer. He argued that the complex polyphonic music of that era obscured the words, making it all but impossible to follow the dramatic storytelling. He advocated for a new musical style that prioritized textual clarity. This led to the development of monody, a solo vocal style accompanied by simple chords, designed to mimic the natural rhythms and inflections of common speech in the contemporary Italian vernacular (not Liturgical Latin). This style allowed singers to deliver lyrics in a way that was immediately intelligible to audiences. The Camerata's experiments culminated in the creation of the first opera, Jacopo Peri's *Dafne*, premiering around 1597 in Florence.[1]

Monody, Recitative, and the Aria: Prioritizing Narrative over Music

The development of monody was a pivotal step in ensuring that opera's narratives were accessible to audiences. Unlike the polyphonic choral music of the 16th century, which layered multiple voices and often obscured the text, monody featured a solo voice supported by a *basso continuo*, where a bass line is played continuously, typically by harpsichord, lute, or chitarrone. This provided a simple harmonic foundation that allowed the singer's voice to remain prominent, ensuring that the lyrics were clear and understandable.[1]

Monody led to the musical style of recitative, which closely followed the accentuation of poetry and imitated the rhythms and accents of spoken language. Recitative enables characters to converse in music, advancing the plot and delivering the narrative without being muddied by musical interference. By reducing the complexity of the accompaniment, composers allowed the voices of the performers to carry the narrative, making the lyrics the focal point of the performance.[1]

Monody also led to the emergence of the aria as a distinct musical form. The aria allowed for moments of emotional reflections sung by virtuosos, as opposed to recitative's role in advancing the plot.

Early Opera Orchestration

The earliest operas used small ensembles to support vocal lines without overpowering them. Ensembles of plucked string instruments and a keyboard, usually a harpsichord or organ, provided harmonic support without obscuring the voices of the singers. As the genre matured, composers experimented with larger and more varied ensembles for dramatic effect.[2]

Claudio Monteverdi's *L'Orfeo* (sometimes referred to as *La Favola d'Orfeo*) premiered in Mantua in 1607. His score calls for an ensemble of approximately 41 instruments which included strings, brass, harpsichords, organs, harp, and recorders. He assigned specific instrumental groups to different scenes and characters to reflect dramatic context.[2]

This strategic use of instrumentation not only distinguished between different locations within the story but also heightened emotional impact. Strings and recorders in pastoral settings reflected the celebratory mood of scenes involving nymphs and shepherds, while heavy brass in the underworld scenes amplified the narrative's emotional weight. Monteverdi advised his musicians to use their judgment to play the music without flourishes or runs. He tells them this would lead to chaos and confusion which would offend the listener and obscure the text.[2]

The First Opera Houses

Based on their design, one can infer that early opera houses likely created a less focused acoustic environment compared to later opera houses. The Teatro Farnese was the first venue specifically built for opera.

The Teatro Farnese was built in Parma, Italy, between 1618 and 1628. It was built in the form of an elongated amphitheater, which included a central arena in front of the stage, and audience seating on three sides of the arena. The arena and stage were used together for performances with dramatic effects like flooding for mock sea battles.[3] This design forces the vocals and music to diffuse across a wide area. When standing in the arena portion, no matter which way performers turned, they would always be facing away from a good portion of the audience. Intelligibility would require precise balancing of the singers and musicians who carefully controlled their dynamics.

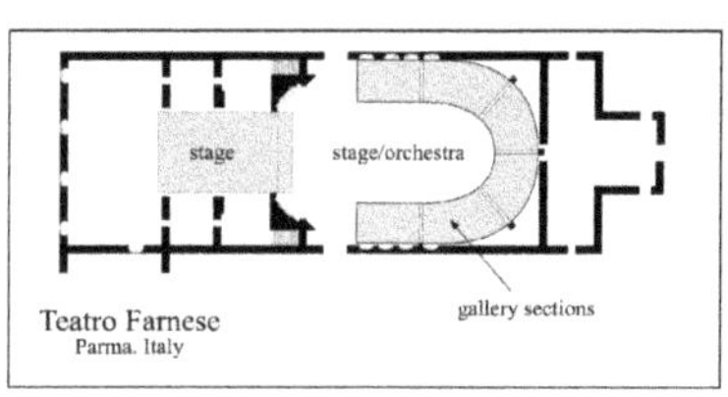

Figure 12.1 Block diagram of Teatro Farnese, Parma, Italy.

Drawing by the author.

There were early opera performances and venues that predate the Teatro Farnese, but they were not purpose-built opera houses. Peri's *Euridice* was performed for the wedding of Maria de' Medici and King Henri IV of France in the theatre in the Pitti Palace in Florence in 1600, which was adapted for the performance. Teatro Mediceo in the Uffizi Palace in Florence was also adapted to host early operas.[3]

Sound Effects in Early Opera

Sound effects often complemented the orchestration. Effects such as thunder, wind, and other natural sounds were carefully timed to occur during instrumental interludes or moments without singing to ensure that they did not interfere with the vocal lines. Instrumental music was often played during the sound effects to mask the unwanted noise that the machinery created.

Chapter 12 Exercises

Exercise 1: Orchestrating for a Modern Musical With Renaissance Principles

Objective: Apply the orchestration and composition techniques of early opera to create a sound design excerpt for a modern musical, emphasizing performer intelligibility and emotional impact.

Instructions:

1. Choose a short scene from a modern musical. Identify a moment in a song where emotional intensity or narrative clarity is critical for the advancement of the plot.
2. Get familiar with the sheet music for the scene.
3. Write a reflection on any danger areas where the music might overpower or muddy up the lyrics. Suggest instrumentation that would support the actor's vocals without pulling focus.
4. Add one sound effect during an instrumental interlude, ensuring it enhances the drama without interfering with the vocals, as described in early opera practices.

Exercise 2: Design a Plan to Overcome the Constraints of Teatro Farnese

Objective: Evaluate the advantages and constraints of Teatro Farnese's ground plan and come up with a plan to both use those advantages to your favor and overcome any constraints.

Instructions:

1. Draw a rough ground plan of Teatro Farnese.
2. Make a list of acoustical advantages.
3. Make a list of acoustical constraints.
4. With your main goals being vocal intelligibility and meaning, submit a proposal on how you would use its advantages and overcome its constraints to deliver a meaningful aural experience to a full audience.

13

The Rise of American Theatre

Theatre in the American colonies was highly influenced by European theatre practices, particularly English traditions. English theatre traditions such as the use of proscenium stages and mechanical sound effects shaped American productions, setting the stage for performances and venues that prioritized vocal and musical propagation and intelligibility, and emotional resonance.

Due to censorship laws of the time, theatres relied heavily on spectacle and special effects more than dialogue-driven plays with broad themes. Sound effect machines like thunder runs and wind machines, inspired by English designs, were adapted creatively in America.

Censorship laws suppressed theatre development, stifling advancements in architectural acoustics. At first, theatre performances were held in temporary structures or borrowed buildings. Over time, as laws began to change, permanent theatres were built, and theatre thrived in major cities such as Philadelphia, New York, and Boston.[1]

Puritan and Quaker Resistance

Initially, theatre in America was met with heavy resistance from Puritan and Quaker communities, particularly in New England, who considered the practice a source of moral corruption and, therefore, incompatible with their religious principles. These communities considered theatre a distraction from their spiritual duties. In 1789, theatre was permitted under strict censorship

DOI: 10.4324/9781003519393-16

laws. These laws delayed the establishment of permanent theatres in cities like Philadelphia and New York, forcing theatre companies like The American Company to seek permissions or to operate covertly. This resulted in limited theatre growth. These laws pushed performances to less regulated areas like Virginia or Maryland.[1]

Theatre companies were adapted by rebranding performances, seeking influential patrons or operating in temporary venues. Gradually, by the late 18th century, theatres gained legitimacy through advocating for its moral utility, which led to the establishment of permanent venues in Philadelphia and elsewhere.[1]

Theatre Design and Acoustics

The first building in America constructed specifically for theatrical performances was the Williamsburg Playhouse, built in Williamsburg, Virginia, in 1716. It was initiated by William Levingston, a merchant, along with Charles and Mary Stagg, London-trained dancing masters, who managed the theatre.[2] The likely wooden construction would have provided a pleasingly resonant environment. The hard surfaces would have provided clear dialogue and music to the audience area.

In 1753, the first theatre building in New York City was constructed on Nassau Street as a home for The London Company of Comedians, which later became known as The American Company.[1]

The Chestnut Street Theatre in Philadelphia, designed by Benjamin Henry Latrobe, opened in 1794. Latrobe was an avid admirer of British theatrical architecture and modeled the Chestnut Street Theatre after the Theatre Royal in Bath, England. It closely followed the Bath theatre's neoclassical style and interior proportions.[3] It featured a classic amphitheater arrangement with tiered boxes and accommodated up to 2,000 patrons. It integrated a dedicated orchestra pit for musicians or choruses, depending on the program, along with green rooms for musicians or actors. The venue hosted elaborate seasons blending spoken plays, operas, choruses, and pantomimes with original scores and overtures.[1]

Although there is not much literature about the acoustics of the space, a few things can be inferred. The amphitheater layout suggests an optimization for natural projection and good evenness in sound distribution across the venue, which is crucial for spoken dialogue plays, as well as for orchestral music in opera. The integration of an orchestra pit and green rooms indicates the space supported complex aural elements from which we can infer low reflection issues from architectural surfaces and sufficient volume for voices and instruments to reach all seats.

The Walnut Street Theatre was built in Philadelphia in 1811. The main stage is arranged in orchestra and mezzanine levels, seating up to 1,054 audience members. It originally had a dome that was removed in 1820 to enhance sound quality, suggesting that it initially caused acoustic issues like echoes. The original construction used brick, with a marble façade added later. These materials likely reflected sound well and probably caused some echoes.

Actor Intelligibility and Clarity

The bulk of the work of vocal clarity fell to the actors' technique. They relied on their training to project their voices across large and often reverberant spaces. Time-tested techniques included diaphragmatic breathing to produce a powerful, sustained sound. Actors trained in precise articulation to ensure audience comprehension.[4]

Strategic stage positioning in performance blocking was crucial to intelligibility. The style of acting was often presentational, with grand gestures and exaggerated vocal delivery. English actor and manager David Garrick, a significant influence on American Theatre, promoted a more naturalistic style of acting, while maintaining robust vocal projection in order to reach the back of the house.[4]

In the 19th century, as theatres grew larger and melodrama evolved, the demand for vocal clarity became even more pressing. Actors performing in venues like the Academy of Music had to maintain intelligibility for audiences seated far from the stage, and so we can expect that they relied mostly on their own vocal training,[5] as acoustic properties were undoubtedly inconsistent from venue to venue.

Sound Effects in Early American Theatre

The machines used for sound effects, inspired by traditional English designs, became even more innovative during this time period. Censorship laws limited the plays that could be produced. This pushed theatres toward spectacle-driven productions that sound effect machines could provide.[4]

A Thunder Sheet was a large, thin copper sheet suspended by a chain. It was shaken to create a rumbling type of distant thunder. The stagehand could vary the volume, shaking less intensely for distant thunder and more intensely as the storm came closer.[4]

A Thunder Barrel/Mustard Bowl mimicked the sound of rumbling thunder as well. Heavy balls or stones were rolled inside a wood or metal barrel-type container (resembling a mustard-grinding bowl).[4]

A Thunder-Run used cannonballs or stones that were rolled down a wooden track. It was often placed above the theatre or backstage to simulate the sound of thunder with a legato type of resonant rumble.[4]

A Wind Machine was made up of two circular frames covered with tightly stretched silk or canvas, each on a spit type frame that could be rotated. The two were spun against each other.[4]

A Wave Machine mimicked the sound of water waves. It was an octagonal box with internal shelves, filled with peas, light rocks, or shells. When the stagehand rotated the box, the contents would bounce around inside to create the sound.[4]

A Rain Machine was somewhat like the wave machine. A large and long four-sided tube with shelves on the inside and filled with small pebbles or beads simulated the sound of rain when tilted from horizontal to vertical.[4]

A Hail Machine was an even larger tube with shelves on the inside and filled with beads, creating the sound of a hailstorm.[4]

A Mechanical Drum was used for creating the sound of explosions. A large parchment skin stretched over a circular frame, much like modern tympani, was struck with a sponge on a whalebone spring. This produced sounds ranging from a near explosion to a distant gunshot. The sponge on the spring created realistic echoes.[4]

Recorded Sound Effects

The late 19th century saw the beginning of technological innovation with the invention of the phonograph in 1877. In 1890, The Terry's Theatre in London's West End was the first theatre to use a recording for a sound effect. The play was called "The Judge," and a phonograph was used to play a recording of a baby crying. In the program for the show, it was announced as the first use of the phonograph for stage purposes.[4]

This event marked a significant global milestone as the expansion of the recording and electronics industry continues to revolutionize the way we create sound to this day.

Chapter 13 Exercises

Exercise 1: Sound Effect Design Challenge

Objective: Understand and recreate early American theatre sound effects using modern materials, connecting historical techniques to contemporary creativity.

Instructions:

1. **Research and Select**: Choose one sound effect machine described in Chapter 13 (e.g., thunder sheet, wind machine, wave machine, or rain machine).
2. **Modern Recreation**: Using everyday materials, design and build a small-scale working model of the chosen sound effect machine. Ensure it can produce a sound resembling the intended effect.
3. **Performance and Reflection**: Demonstrate your device to the class, explaining how it mimics the historical machine.

Exercise 2: Live Sound Plot Design for a Contemporary Play

Objective: Design a sound plot for a scene from a contemporary play, using only live, non-recorded sound effects inspired by early American theatre techniques, to enhance the scene's atmosphere and emotional impact.

Instructions:

1. **Select a Scene**: Choose a scene from a contemporary play or musical.
2. **Design the Sound Plot**: Create a sound plot that uses only live, non-recorded sound effects. For each of the three moments:
   - Specify the sound needed.
   - Describe a live method to produce it, adapting historical techniques with modern materials.
   - Explain how a stagehand would operate the device (e.g., timing, intensity, or positioning backstage).
   - Ensure the method does not use any recorded or electronic sound.
3. **Presentation**: Demonstrate one of your live sound effects using the materials described. Explain how it fits the scene and enhances the storytelling.
4. **Submission**: Submit the written sound plot document via the course platform.

Glossary

Acoustic Artifact: An unintended or undesirable sound introduced during audio recording, processing, or playback.

Acoustical Node: (Or simply "node" in acoustics) is a point along a standing wave where the amplitude of vibration or displacement is zero (or at its minimum), resulting from destructive interference between incoming and reflected waves. These nodes are fixed points that do not move during the wave's oscillation, often occurring at boundaries like the ends of a vibrating string or in room modes where sound waves cancel out. Nodes can cause uneven sound distribution and "dead spots" where bass frequencies are absorbed.

Acoustic Reflection: Sound waves bouncing off theater surfaces, creating secondary auditory effects that impact clarity.

Acoustical Constraint: Environmental factors, like excessive reverberation or noise, that hinder optimal sound projection.

Acoustical Focus: Architectural features, such as curved walls, directing sound to a specific point in a room.

Acoustics: The scientific study of sound wave propagation, reflection, and absorption, especially in architecture.

Actor Intelligibility: The clarity with which an actor's dialogue or lyrics are understood by the audience.

Actor Projection: Trained vocal techniques used by actors to deliver dialogue audibly to distant audience members. Projection is more about dynamics traveling as far as possible and not only about volume.

Aesthetic: A personal sense of artistic beauty shaped by cultural and individual influences.

Alarum: A loud auditory signal, like drum rolls or trumpet blasts, indicating urgent action in historical dramas.

Ambient Sound: Background noises, such as environmental sounds. Ambient sound effects can add meaning to a scene.

Amphitheater: An open-air theater with a central stage and tiered seating for natural acoustic projection.

Amplification: An increase in sound intensity, generally perceived as volume.

Aria: A solo vocal piece in opera expressing deep emotion, distinct from narrative-driven recitative.

Art: The deliberate creation of works that evoke emotional and intellectual responses. In sound design theory, art is defined as *The expression of a meaningful idea.*

Artistic Expression: An artist's unique creative output reflecting personal vision and emotional significance. That expression can take the form of literature, speech, song, music, painting, dance, theatre, etc.

Atmospheric Sound: Background sound cues that establish location, time period, or mood, enhancing the immersive quality of the theatrical environment.

Audience Perception: How spectators interpret auditory stimuli based on personal and cultural contexts.

Auditory Experience: The complete sonic environment designed to shape audience interaction with a performance.

Auditory Illusion: Sound effects that simulate events or spaces not physically present on stage.

Aural Atmosphere: The sonic landscape, including music and effects, setting a performance's emotional tone.

Balance: The adjustment of audio levels to harmonize dialogue, music, and effects used for narrative clarity.

Barrel Organ: A mechanical instrument playing preset tunes for atmospheric music in early theaters.

Basso Continuo: A continuous bass line, often by harpsichord, supporting vocal clarity in early opera.

Bellows: A device generating compressed air, often used for historical sound effect mechanisms.

Blocking: Choreographed actor movements ensuring narrative coherence and effective sound projection.

Catharsis: Emotional purification achieved through audience empathy with dramatic characters.

Chambers: Often used in the Renaissance Era to refer to cannons or specific components of cannons due to the design and function of early artillery. Theatres used them for simulating explosive gunfire in dramatic sequences.

Chitarrone: A long-necked string instrument used for rich, resonant tones.

Choices: Deliberate decisions made to express an artistic intention.

Clapper: A wooden device struck to produce sharp sounds mimicking gunfire or impacts.

Collaboration: The integration of expertise from professionals to achieve a unified artistic vision.

Conditioning Factor: Media exposure shaping audience expectations and responses to performances.

Constraint: Practical or creative limitations, such as budget, time, or identity, that guide and enhance choices by defining the rules within which artistic expression must operate.

Context: Referring to sound cues, context is the circumstances surrounding a sound cue, including dialogue, action, setting, and temporal dynamics, that uniquely shape its emotional and narrative meaning in a theatrical production.

Creativity: The synthesis of existing ideas into novel artistic ideas or concepts.

Cue Sheet: A document listing the sequence of sound cues for technical execution during performances. It lists cue number, name, how to execute the cue, and when.

Cue Timing: The precise synchronization of auditory cues with stage action for dramatic impact.

Dance of Satyrs: Lively drum and pipe melodies with percussive stomping for a festive atmosphere.

Dead March: A slow, muted drum cadence marking solemn or tragic moments in performances.

Delivery system: In sound design, this is the method or technique used to provide sound to an audience, such as a public address system or headphones.

Designed Sound Effects: Artificial sound cues of unnatural or unrecordable phenomena, such as spacecraft or futuristic weapons, crafted to fit the narrative.

Diaphragmatic Breathing: Deep, controlled breathing used by actors to sustain vocal projection.

Diegetic Sound: Sounds originating within the narrative world, like a character's radio, enhancing realism.

Digital Audio Workstation (DAW): Software for recording, editing, and mixing computer audio files.

Dispersion: The directing or dispersal of sound waves in many or all directions.

Dramatic Effect: In sound design, sound or context choices amplifying the emotional and narrative impact of a scene.

Dramatic Sound or Scenic Sound: Sound cues integral to the story, occurring within or between scenes to advance the narrative or enhance its emotional impact.

Dynamic Processing: Audio manipulation, like compression, refining sound dynamics for clarity and impact.

Dynamic Range: The range of amplitude or volume levels in a specific measure of time, such as a performance; OR the range of frequencies in a particular sound. Equalization processing manipulates dynamic range, for example.

Echo: Sound waves reflected creating repeated auditory after effects.

Emotional Resonance: In sound design, fostering deep emotional connections between audience and elements such as sound or narrative.

Environmental Sound Effects: Sounds simulating natural phenomena, like wind.

Equalization (EQ): The adjustment of frequency bands to enhance sound clarity or stylistic effects.

Expressionism: A theatrical style prioritizing subjective emotional expression over realistic representation.

Feedback: An audio loop from microphone to speaker, causing disruptive screeches.

First Principles: In sound design, fundamental truths guiding the design and execution of theatrical sound.

Florentine Camerata: A Renaissance group pioneering opera by reviving Greek tragic forms through music.

Foley: The art and technique of creating and recording sound effects in post-production to enhance the audio of films, videos, or other media.

Frequency: Rate at which a sound wave vibrates, measured in Hertz (Hz)

Frequency Band: A specific range of frequencies.

Gallery: In theatre venues, the seating or musician platforms.

groundlings: Standing spectators in amphitheater yards in the Renaissance period.

Hail Machine: A tube with beads tilted to replicate the sharp sound of a hailstorm.

Hall-Shaped Playhouse: A rectangular indoor theater focusing sound on the audience.

Harmonic Scale: In audio and music theory, refers to a sequence of frequencies that are integer multiples of a fundamental frequency, known as *harmonics* or *overtones.*

Hautboy: A reedy wind instrument producing elegant or eerie tones for specific scenes.

Heavens: In Renaissance theatre, the stage roof housing representing *Heaven*, employing mechanisms for effects actor descents.

Hell: In Renaissance theatre, the under-stage area representing the demonic realm, accessed often via trapdoors.

Hertz (HZ): the unit of measurement for frequency, representing the number of cycles per second in a periodic phenomenon, such as a sound wave or vibration.

Holistic Choice: In sound design, decision integrating diverse experiences and emotional insights for a cohesive artistic impact.

House Sound: Sound cues played before, during, or after a performance, such as pre-show music, entr'acte, curtain call, or walkout sound, to set or sustain the audience's mood.

Humanism: Originating in the Renaissance, a philosophy emphasizing individual potential through scholarly education.

Imaging: In sound design, the effect of the location where a sound originates.

Immersive Sound: Enveloping auditory elements integrating audiences into the narrative's sonic world.

Instrumental Music: Non-vocal compositions.

Intelligibility: In sound design, the degree in which a performer's vocals are clearly understood to any given audience member.

Intention: In sound design, a deliberate artistic purpose guiding sound design to align with narrative goals.

Lavalier Microphone: A small, clip-on microphone amplifying actor voices discreetly.

Live Effects: Manually produced sounds executed during a performance for immediate impact.

Live Sound Cues: Real or performed sounds executed on cue during a performance, such as live music, actor-generated sounds, or effects like a crash box or thunder sheet.

Masque Music: The musical compositions associated with the masque, a form of courtly entertainment that flourished in England during the 16th and 17th centuries.

Mastery: Expertise gained through sustained, disciplined practice in a craft.

Meaning: The quality of a sound design that inspires an emotional response in the audience, shaped by individual interpretation and context, not dictated by the designer's intent.

Melodrama: A genre using exaggerated emotions and bold sound effects to engage audiences.

Minimalism: In theatre, a theatrical style using sparse elements to convey profound meaning or emotion.

Mirror Mechanism: Neurological processes enabling empathy through vocal or visual cues.

Mixing Engineer: A specialist adjusting audio levels dynamically for optimal live performance clarity.

Monody: A solo vocal style with minimal accompaniment, prioritizing lyrical clarity in opera.

Music Gallery: An elevated platform for musicians ensuring optimal sound projection in theaters.

Musical Transition: Musical segments bridging scenes to maintain continuity and mood.

Mustard Bowl: A barrel with rolling stones creating resonant thunder effects in performances.

Narrative Arc: The structured progression of a story.

Narrative Clarity: The clear understanding of a story's progression.

Naturalistic Acting: Subtle, realistic performances prioritizing authenticity in theater.

Noise: In sound design, any auditory element, such as backstage clatter or external traffic, that the sound designer does not intend the audience to hear.

Noise Floor: The level of background noise in a sound system or environment, measured in decibels (dB), that represents the lowest possible signal level before it is masked by inherent noise from equipment, environment, or recording medium.

Non-diegetic Sound: External sounds, like background music, shaping mood outside the narrative.

Offstage Sound: Sounds from backstage implying actions or events beyond the visible stage.

Ok plateau: A stagnation in skill development due to a lack of deliberate improvement.

Opera: A genre combining music, drama, and dance for emotionally compelling narratives.

Orchestra: An ensemble of musicians supporting vocalists or amplifying dramatic intensity.

Orchestration: A musical arrangement played by an orchestra.

Panning: The amplitude relationship of sound across speakers for dynamic spatial auditory effects. Stereo panning would involve two channels, a left and a right.

Paperwork: Technical documents, like cue sheets, aiding communication, memory, and design archival.

Patch Sheet: A diagram showing audio equipment connections for a sound system.

Pavane: A slow, stately Renaissance dance accompanied by formal music.

Perspective: In sound design, the chosen point of view for a sound cue, reflecting how a character or the audience perceives it.

Playback of Recorded Sound Cues: Pre-recorded sounds, such as effects or music, played back on cue via media platforms like tape, CDs or computer files to support the narrative.

Polyphonic Music: Complex, multilayered musical compositions with independent melodies.

Presentational Acting: Exaggerated performances engaging audiences through bold delivery.

Production Constraints: Limitations, like budget or space, affecting theatrical design choices.

Professionalism: The practice of living and working according to personal values, such as mastery of skill, free expression, and collaboration.

Prop Sound: Sounds from physical objects, like clashing swords, enhancing narrative realism.

Proscenium Stage: A framed stage focusing audience attention, separating performers from spectators.

Psychoacoustics: The scientific study of how humans perceive, process, and interpret sound.

Rain Machine: A tube with pebbles tilted to simulate the sound of falling rain.

Recitative: A speech-like musical form in opera advancing the narrative with clear lyrics.

Resonance: The amplification of sound through vibrational interaction with a medium.

Resonant Chamber: A space designed to amplify specific sound frequencies for richer acoustics. This could include the resonating chamber of a musical instrument, like a cello, or the acoustical qualities of a given room or venue.

Reverberation: The persistence of sound in a space after the original sound source has stopped, caused by multiple reflections of sound waves off surfaces like walls, ceilings, and floors.

Ritualistic Sound: Sounds, like chants or bells, underscoring ceremonial or spiritual moments.

Satyr: A creature from Greek mythology that is half man and half goat, often associated with Dionysus. Satyrs appeared in ancient drama, especially *satyr plays*, symbolizing revelry, nature, and the uninhibited aspects of humanity.

Scaenae Frons: An ornate backdrop in classical theaters enhancing acoustic and aesthetic properties.

Scene: A narrative segment advancing the story through specific actions or dialogue.

Sennet: A trumpet fanfare announcing royal processions with authority and grandeur.

Show Control: Automated systems synchronizing sound, lighting, and stage effects precisely.

Signal Flow Diagram: A chart tracing the path of audio signals through a sound system.

Sound: Intentional auditory elements, such as dialogue, music, or sound effects, designed to enhance the narrative and emotional impact of a theatrical performance.

Sound Absurdism: Sound cues that are illogical or inappropriate to the context, such as an elephant trumpet in an urban setting, to evoke disorientation or challenge conscious reasoning.

Sound Cue: A timed auditory element supporting stage action and narrative progression.

Sound Design: The planning and execution of auditory elements shaping the theatrical experience.

Sound Design Theory: A framework of fundamental principles guiding the sound designer's artistic process, standardizing understanding and enabling the creation of meaningful, complex auditory experiences.

Sound Effect: A sound produced to represent the sound of something else, created by other means, that is, striking a woodblock to represent a blow to the head.

Sound Expressionism: Sound cues that exaggerate emotional qualities over realism, amplifying a single truthful characteristic to convey a character's inner experience.

Sound Plot: A numbered list of sound cues with instructions for execution during a show.

Sound Realism: Sound cues that closely mimic real-world sounds, delivered through live, recorded, or effect-based means.

Sound Reinforcement: The use of audio systems to amplify and clarify live sound.

Sound Surrealism: Bizarre or illogical sound cues designed to evoke subconscious meaning, creating dreamlike or disorienting effects within the narrative.

Sound Symbolism: Aural metaphors where one sound represents another to convey deeper truths, such as a lion's roar replacing a cat's meow to make a symbolic statement.

Sound System: The collection of audio equipment delivering sound to an audience.

Spatial Effect: Auditory placement creating a three-dimensional sound experience for realism.

Stage Acoustics: The physical properties of a stage influencing sound projection and perception.

Stage Blocking: Choreographed movements aligning with auditory cues for effective storytelling.

Stage Crew: Technical personnel managing sound, lighting, and stage operations during shows.

Stage Direction: Scripted instructions guiding actions, actor placement, effects, or other information for clarity.
Stagehand: A crew member handling technical tasks, like sound or prop management, backstage.
Standing Wave: A standing wave is a sound wave that remains fixed in place within a confined space, like a room, due to repeated reflections off the boundaries, forming a stable pattern of still points (nodes) and vibrating peaks (antinodes) instead of traveling away.
Suspension of Disbelief: Audience acceptance of theatrical illusions for narrative immersion.
Theatrical Convention: Established practices, like fanfares, shaping audience expectations.
Theme: In theatre, a central idea or message explored through a performance's narrative and sound. A musical composition expressing an idea, person, or other element.
Thunder Barrel: A barrel with rolling stones producing deep thunder sounds for dramatic effect.
Thunder Run: A wooden track with rolling cannonballs creating rumbling thunder effects.
Thunder Sheet: A shaken metal sheet simulating rolling thunder for stormy atmospheres.
Tiring-House: A backstage area for actor preparation and offstage sound effect execution.
Tonal Quality: The pitch and timbre of a sound defining its emotional or narrative role.
Transitional Sound: Sound cues that guide the audience into, out of, or between scenes, maintaining narrative momentum or shifting emotional tone.
Trapdoor: A concealed stage opening for dramatic entrances and exits.
Underscore: Non-diegetic music or sound effects, imperceptible to characters, that reflect emotions or alter the context of dialogue or action to propel the narrative.
Venue Acoustics: The sonic characteristics of a performance space affecting sound clarity.
Vocal Mimicry: Actors vocally replicating environmental, animal or other sounds.
Voice-overs: Recorded or live spoken cues, often non-diegetic, revealing a character's thoughts or providing narrative commentary to enhance audience understanding.
Volume: The perceived loudness of a sound cue.
Wave Machine: A rotating container with peas or shells simulating rhythmic water wave sounds.

Whalebone Spring: A flexible rod on a drum producing echoing effects for explosions or shots.

Wind Machine: A rotating frame with silk generating howling or gentle wind sounds.

Woodblock: A struck wooden block producing sharp, percussive sounds for rhythmic effects.

Yard: The open area in amphitheaters, where standing spectators add dynamic energy.

List of Sources by Chapter

Chapter 2: Foundational Concepts in Sound Design Theory

1. Iacoboni, Marco, *Mirroring People: The Science of Empathy and How We Connect with Others* (Farrar, Straus, and Giroux, 2025).
2. Foer, Joshua, *Moonwalking with Einstein: The Art and Science of Remembering Everything* (New York: Penguin Press, 2011).

Chapter 3: Intentions and Choices

1. Greenberg, Richard. *Three Days of Rain* (New York: Dramatists Play Service, 1998).
2. Frankl, Viktor E., *Man's Search for Meaning*, Kindle ed. (Beacon Press, 2006).
3. Covey, Stephen R., *The 7 Habits of Highly Effective People: 30th Anniversary Edition*, Kindle ed. (Simon & Schuster, 2020).

Chapter 4: From Sound to Meaning

1. Rauschecker, J. P., & Tian, B., *Mechanisms and Streams for Processing of "What" and "Where" in Auditory Cortex* (Proceedings of the National Academy of Sciences, 97(22), 11800–11806. 2000).
2. Hohwy, J., *The Predictive Mind* (Oxford University Press, 2013).
3. Barrett, Lisa Feldman, *How Emotions Are Made: The Secret Life of the Brain*, Kindle ed. (Boston: Houghton Mifflin Harcourt, 2017).

Chapter 6: The Sound Designer's Toolbox

1. Sayers, Peig, *Peig: The Autobiography of Peig Sayers of the Great Blasket Island*, 1st American ed., Translated by Bryan MacMahon (Syracuse, NY: Syracuse University Press, 1974).

Chapter 9: Early and Late Classical Period: The Greeks and the Romans

1. Csapo, Eric, & Miller, Margaret C. *The Origins of Theatre in Ancient Greece and Beyond* (Cambridge University Press, 2007).
2. Arnott, Peter D., *Public and Performance in the Greek Theatre* (Routledge, 1991).
3. Lynch, Tosca A. C., & Rocconi, Eleonora, eds. *A Companion to Ancient Greek and Roman Music* (Hoboken, NJ: Wiley-Blackwell, 2020).
4. Aristotle, *Poetics*, Translated by S. H. Butcher, in The Poetics of Aristotle (Project Gutenberg, 2013).
5. David, Bain, *Actors and Audience: A Study of Asides and Related Conventions in Greek Drama* (Oxford Classical and Philosophical Monographs. Oxford: Oxford University Press, 1977).
6. Green, J. R. *Theatre in Ancient Greek Society* (Routledge, 1994).
7. Tsilfidis, Alexandros, et al., *Function and Acoustic Properties of Ancient Greek Theatre Masks* (Acta Acustica united with Acustica, 2013).
8. Vitruvius, *De Architectura*, Translated by Morris H. Morgan (Project Gutenberg, 2006).
9. Valière, J.-C., et al., *Acoustic Pots in Ancient and Medieval Buildings: Literary Analysis of Ancient Texts and Comparison with Recent Observations in French Churches* (Acta Acustica united with Acustica, 2013).
10. Collinson, David, *The Sound of Theatre*, Kindle ed. (Entertainment Technology Press Ltd, 2020).
11. Aristotle, *De Anima*, Translated by D. W. Hamlyn (Clarendon Press, 2002).
12. Permanent Delegation of Greece to UNESCO, *Ancient Greek Theatres* (United Nations Educational, Scientific and Cultural Organization (UNESCO), 2014). https://whc.unesco.org/en/tentativelists/5869

Chapter 10: The Middle Ages

1. Symes, Carol, *The History of Medieval Theatre/Theatre of Medieval History: Dramatic Documents and the Performance of the Past* (History Compass 7, 2009). https://doi.org/10.1111/j.1478-0542.2009.00613.x
2. Enders, Jody ed., *A Cultural History of Theatre in the Middle Ages* (Bloomsbury Academic, 2017).
3. Bevington, David. *Medieval Drama* (Houghton Mifflin, 1975).
4. Twycross, Meg. The Theatricality of Medieval English Plays. in *The Cambridge Companion to Medieval English Theatre*, edited by Richard Beadle (Cambridge University Press, 1994).

5. Tydeman, William. *The Theatre in the Middle Ages* (Cambridge University Press, 1978).
6. Orme, Nicholas, *The Medieval Church: Your Guide to Religion & Worship in the Middle Ages* (History Extra, 2022). https://www.historyextra.com/period/medieval/medieval-church-churches-middle-ages-religion.
7. Berardi, Umberto, et al., *The Evolution of the Acoustics of a Medieval Church* (Canadian Acoustics – Acoustique canadienne, 48(3), 2020).
8. Quintilian, *Institutio Oratoria, Book XI*, Translated by Butler (H.E. Lyceum Institute, 2024).
9. Collinson, David, *The Sound of Theatre*, Kindle ed. (Entertainment Technology Press Ltd, 2020).

Chapter 11: The Renaissance – Transitioning to the Modern Theatre

1. Magagnato, Licisco, *The Genesis of the Teatro Olimpico* (Journal of the Warburg and Courtauld Institutes, 14(¾), 1951).
2. Gurr, Andrew, *The Shakespearean Stage, 1574–1642*, 4th ed. (Cambridge University Press, 2009).
3. Collinson, David, *The Sound of Theatre*, Kindle ed. (Entertainment Technology Press Ltd, 2020).
4. Smith, Bruce R., *The Acoustic World of Early Modern England* (University of Chicago Press, 1999).
5. Lawrence, W. J., *Pre-Restoration Stage Studies* (Harvard University Press, 1927).

Chapter 12: The Italian Renaissance – The Birth of Opera

1. Abbate, Carolyn & Parker, Roger, *A History of Opera: The Last Four Hundred Years* (Penguin Books, 2012).
2. Beat, Janet E., Monteverdi and the Opera Orchestra of His Time. in *The Monteverdi Companion*, edited by Denis Arnold and Nigel Fortune (London: Faber and Faber, 1968).
3. Forsyth, Michael, *Buildings for Music: The Architect, the Musician, and the Listener from the Seventeenth Century to the Present Day* (Cambridge University Press, 1985).

Chapter 13: The Rise of American Theatre

1. Dunlap, William, *A History of the American Theatre from Its Origins to 1832* (University of Illinois Press, 2005).
2. Bullock, Helen. *Colonial Theatre Historical Report, Block 29 Building 17A*. Research Report No. 1586. (Williamsburg, VA: Colonial Williamsburg Foundation, 1935 and Reissued digitally 1990). https://research.colonialwilliamsburg.org/DigitalLibrary/view/index.cfm?doc=ResearchReports\RR1586.xml
3. Hartnoll, Phyllis, *The Theatre, A Concise History* (Thames and Hudson, 1968).
4. Collinson, David, *The Sound of Theatre*, Kindle ed. (Entertainment Technology Press Ltd, 2020).
5. Barron, Michael, *Auditorium Acoustics and Architectural Design*, 2nd ed. (Spon Press, 2009).

Index

Page numbers in *italics* indicate figures; page numbers in **bold** indicate tables.

For Product Safety Concerns and Information please contact our EU representative GPSR@taylorandfrancis.com
Taylor & Francis Verlag GmbH, Kaufingerstraße 24, 80331 München, Germany

www.ingramcontent.com/pod-product-compliance
Ingram Content Group UK Ltd.
Pitfield, Milton Keynes, MK11 3LW, UK
UKHW050058200926
475702UK00019B/2900
* 9 7 8 1 0 3 2 8 5 6 7 8 0 *